Visual Diagnosis Self-tests on

Rheumatology

THIRD EDITION   *The Continuing Medical Education Series*

merit
PUBLISHING
INTERNATIONAL

Acknowledgements

We also wish to express our thanks to our Rheumatology, Internal Medicine, Radiology, and Pathology colleagues at both University of South Florida and James A Haley Veterans Hospital whose comments have helped with optimizing this book:

Frank Vasey MD

Colleen Ward MD

Jason Guthrie MD

Natalie Brown MD

Aasim Rehman MD

Priya Reddy MD

Jennifer Reid MD

Gregory Carney MD

Jose Lezama MD

Vanessa Osting MD

Kimberly Harding

Contents

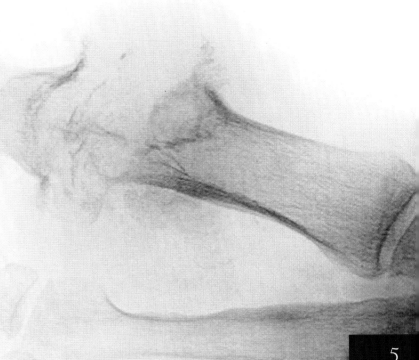

Foreword

Clinical rheumatology is a complex and rapidly advancing field. Considerable progress has occurred in recent years on several aspects, especially concerning pathogenesis, imaging studies, and therapy. Trying to keep up with recent advances in the field is highly demanding, although there is considerable information disseminated through clinical symposia and seminars, medical journals, textbooks of the specialty, videos, databases and other pertinent literature.

Dr. Valeriano-Marcet and her colleagues from the Division of Rheumatology at the University of South Florida, College of Medicine, Tampa, Florida, have put together this book, which nicely highlights and emphasizes a practical and evidence-based approach to the management of common and not so common rheumatic disorders as they occur in daily clinical practice.

A total of 27 cases are presented and fully discussed by a team of experienced authors. These cases encompass patients with inflammatory articular disorders such as rheumatoid arthritis and spondyloarthritides, metabolic disorders such as gout and Gaucher's disease, vasculitides, connective tissue disorders such as lupus, dermatomyositis, and polymyositis, and degenerative disorders. The information presented is concise, up-to-date, well illustrated, and provides appropriate guidance on the management of a wide scope of rheumatic disorders, laboratory evaluation, differential diagnosis, and therapy.

"Visual Diagnosis in Rheumatology" is a very good addition to our rheumatologic literature. Medical students, medical residents, rheumatology fellows, family physicians, internists, clinical rheumatologists, and other health related personnel alike who treat patients afflicted with these disorders should find this text to be an invaluable resource in the management of rheumatic disorders.

Luis R. Espinoza, M.D.
Professor and Chief
Section of Rheumatology
LSU Health Sciences Center
New Orleans, LA 70012-2822

Introduction

Rheumatology encompasses a wide variety of complex musculoskeletal conditions, which often present with joint and muscle complaints. Many systemic rheumatic diseases also have a multitude of extra-articular manifestations. It is important to perform a comprehensive history and physical examination to help recognize unusual features.

The goal of this book is to illustrate the multitude of clinical findings as well as important radiologic and pathologic findings associated with rheumatic diseases. A multi-disciplinary approach is important for optimal patient care.

We have included both common and rare conditions to help review important aspects of our very interesting specialty of Rheumatology.

We hope that this updated edition will serve as a review as well as stimulate interest to search for further information, and thus we have included several references pertaining to each case. Our questions have been designed to summarize the recent changes and updates particularly in therapeutics.

This is a very exciting time in Rheumatology as recent advances are changing the pattern of disease and providing much needed relief for our patients.

Case 1

A 45-year-old man has approximately 20 years of back pain throughout the spine but worst in the low and mid back. He has exercised regularly but has noticed increased difficulty bending forward. He can no longer tie his shoes; in fact, he can barely reach his hands to his knees when leaning forward. He is worst first thing in the morning when getting up from bed. He has two to three hours of morning stiffness. A hot shower provides some relief. He has tried multiple non steroidal ant-inflammatory agents (NSAIDs) over the years but they have provided little benefit in the last year or two. His father had similar back problems.

On examination the patient has significant loss of the usual lumbar lordosis and very limited flexion of the spine. He has no tenderness over the SI Joints.

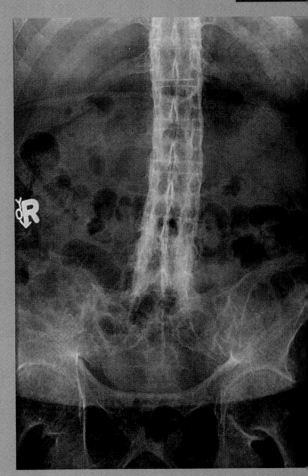

Figure 1. Anterior projection of AP spine.

1. What are the radiographic findings and what is your diagnosis?

2. How do the radiographic features differ between Ankylosing Spondylitis (AS) and spondylosis?

3. What is the difference between back pain from degenerative disc disease and AS of the spine?

4. What are other manifestations of AS?

Answers to Case 1

Findings

This is a typical presentation of ankylosing spondylitis (AS). The X-ray shows a classic "bamboo spine" with syndesmophytes – ossification involving the annulus fibrosis bridging adjacent vertebrae. These features lead to a rigid spine with limited flexibility and increased fragility. Obliteration of the sacroiliac joints is also represented on this X-ray.

Radiographic features

In contrast to AS, the radiographic features of spondylosis (degenerative disc disease) include degenerative spurs or osteophytes, which extend at right angles to the vertebrae rather than following the contour of the disc. There is disc space narrowing in spondylosis. There is no sacroiliitis and usually no extra articular manifestations.

Differences

AS is an inflammatory arthritis with the classic feature of being more painful in the morning and less painful with activity, and morning stiffness of greater than one hour. Mechanical back pain due to spondylosis usually worsens after activity, and is associated with minimal morning stiffness.

Other Manifestations

Osteoporosis may develop secondary to the lack of movement and usual re-modeling of the skeleton. Often there is complete loss of range of motion of the neck. Bilateral sacroiliitis may also occur. Chest expansion is often limited. Uveitis and apical pulmonary fibrosis may be other manifestations, as well as aortic regurgitation, amyloidosis and IgA nephropathy.

Case 2

A 65-year-old man with a 15-year history of seropositive, nodular rheumatoid arthritis presents with a three day history of the sudden onset of numbness of the feet. He comes in today because he has noted difficulty ambulating. The patient's arthritic symptoms have been well controlled on prednisone (5 mg daily) and methotrexate (15 mg weekly).

Physical examination revealed a chronically ill appearing man in no apparent distress. Musculoskeletal examination demonstrated chronic deformities of the hands with ulnar deviation, and subluxations of the MCP joints. There was proliferative synovial swelling of the wrists with decreased ROM, but no tenderness. There were bilateral olecranon nodules. There was no lumbar spinal tenderness and bilateral straight leg raising test was negative. His neurological examination was remarkable for decreased sensation over the dorsum of the foot and left lateral lower leg, inability to dorsiflex the L foot, and absent left ankle reflex.

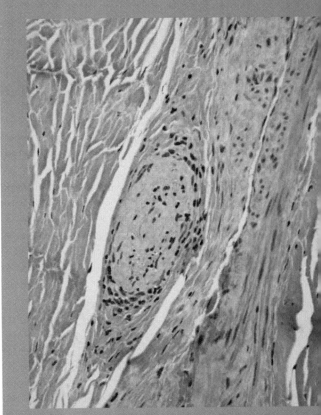

Figure 2. *Sural nerve biopsy.*

Pertinent laboratory results: Hgb 12.6, ESR 95 mm/hr, SPEP consistent with polyclonal hypergammaglobulinemia.

EMG and NCS reveal a distal, asymmetric sensorimotor axonal polyneuropathy.

1. **What does the biopsy show?**

2. **What is your diagnosis?**

3. **What subset of patients is at risk for this complication?**

4. **What are the clinical features?**

5. **How is the diagnosis confirmed?**

Answers to Case 2

Biopsy results

A medium size artery with intimal proliferation and luminal occlusion.

Diagnosis

Rheumatoid vasculitis (RV)

Population at risk

The condition occurs in a minority of patients with rheumatoid arthritis. It tends to occur in men with longstanding, seropositive, destructive rheumatoid arthritis. Most patients have rheumatoid nodules. The vasculitis tends to occur at a time when the active erosive process in the joints has become less active. "The vasculitis chases the arthritis away."

Clinical features

RV may affect a wide range of blood vessel types, from medium-sized muscular arteries to somewhat smaller arterioles to post-capillary venules. Within a given patient, clinical features of both medium- and small-vessel disease may be found.

The areas of the body involved most commonly by RV are the skin, digits, peripheral nerves, eyes, and heart. Cutaneous vasculitis, the most common manifestation, is found in up to 90% of patients with rheumatoid vasculitis, and presents with deep LE ulcers, digital ischemia, necrosis, and gangrene when medium sized vessels are affected, and palpapable purpura when small vessel vasculitis is present.

Neurologic involvement presents with a sensory neuropathy in 40%, and mixed sensory motor neuropathy in 20% of patients. Mononeuritis multiplex results from infarction of individual peripheral nerves due to vasculitis in the vasa nervorum. Mononeuritis multiplex typically affects distal nerves in an asymmetric and asynchronous pattern. The onset is rapid, with sensory changes, most commonly numbness, followed within days to weeks by loss of motor function. Recovery from severe vasculitic neuropathy takes from 12 to 18 months, and may not be complete, even with immunosuppressive therapy.

Ocular manifestations of RV include scleritis, episcleritis, and peripheral ulcerative keratitis (PUK).

Cardiac manifestations of RV include pericarditis, and coronary arteritis. Coronary arteritis leading to myocardial infarction in RV is extremely rare.

Confirming the diagnosis

Since treatment of RV requires substantial immunosuppression and potential toxicity, whenever possible, tissue should be obtained. Potential sites for biopsy include skin, sural nerve and muscle, and renal biopsy if active urinary sediment or renal dysfunction is present. Angiography is rarely useful in RV since the size of the vessels involved is lower than the resolution of the test; additionally findings are often nonspecific in RV.

Case 3

An 80-year-old female complained of recurrent swelling of her toes, elbow, fingers and occasionally knees. These episodes previously lasted for seven to ten days but now they are more difficult to tolerate as they last up to three weeks. There is usually severe pain with redness and it is difficult to walk. She is getting frustrated with the episodes occurring more frequently. Antibiotics were given for possible cellulitis.

The patient denied any trauma or infections in the last few weeks. She recently increased her water pill to try and help with the swelling. She had kidney stones when younger, but no problems in the last ten years. Her diet included seafood almost every day, with one or two beers a night. She remembered her father having similar problems with his feet.

Examination revealed swelling of both great toes with erythema and exquisite tenderness to touch. The erythema extended to the dorsum of the right foot and ankle. She had decreased range of motion of all toes and her ankle range of motion was very limited. She had several nodules on the helices of her ears.

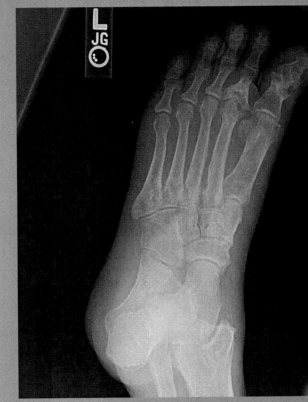

Figure 3.

Labs reveal BUN 25, Cr 1.9.

Synovial fluid aspirate of the right ankle under polarizing microscope revealed negatively birefringent intra- cellular needle shaped crystals.

X-rays of the foot are shown revealing the typical over-hanging edges of large erosions at the first IP joint and erosive changes at the second MTP

1. **What are the predisposing factors increasing the likelihood of gout?**

2. **How would you treat the acute symptoms of gout in this patient?**

3. **What are the indications for allopurinol and when should it be started?**

Answers to Case 3

Predisposing factors

Many factors in combination predispose to the likelihood of gout attacks and include: Increasing age, renal insufficiency, diurectic use, a family history of gout and high purine diet.

Treatment of acute symptoms

Acute treatment options in general for gout include oral colchicine, NSAIDs, Adrenocorticotropic hormone (ACTH) and oral or intrarticular steroids. In this patient NSAIDs were avoided because of renal insufficiency and oral prednisone was chosen for treatment of this polyarticular presentation.

Treatment

Therapy with allopurinol is indicated for recurrent gouty attacks, tophi, joint damage on X-rays, renal stones and chemoprophylaxis for the tumor lysis syndrome. There is no indication for use of allopurinol for asymptomatic hyperuricemia. Allopurinol should be delayed for several weeks after an acute attack has resolved to prevent a further exacerbation. It should be started at a low dose and adjusted for renal insufficiency, to prevent occurrence of the allopurinol hypersensitivity syndrome. It is important to remember that soft tissue swelling related to a gouty exacerbation can mimic cellulitis.

Case 4

A 56-year-old man was well until four months ago when he developed the sudden onset of proximal upper and lower extremity myalgias. He was placed on prednisone 15 mg with some improvement, and one month later this was increased to 20 mg. Despite this treatment, he experienced persistent fatigue, myalgias, and a 15 pound weight loss. He has recently noted a burning sensation in the soles of his feet. Medications include prednisone (20 mg daily) and omeprazole (20 mg daily).

Physical examination revealed a chronically ill appearing gentleman with cushingoid facies. Pulse 80 Blood Pressure 170/95. There was a mild livedo patterning present on examination of the skin. There were multiple tender violaceous nodules (0.5 to 1 cm) clustered over the medial aspect of the knees and lateral soles (Figure 4). Mild proximal upper and lower extremity tenderness was present. The right first, and fifth and left first, fourth, and fifth toes were cyanotic (Figure 5). Femoral, Popliteal, Dorsalis Pedis and Posterior tibial pulses were intact. Neurological examination revealed decreased strength of dorsiflexion of the left foot, and decreased light touch in the distal toes.

Figure 4.

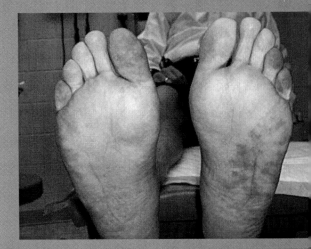

Figure 5.

25

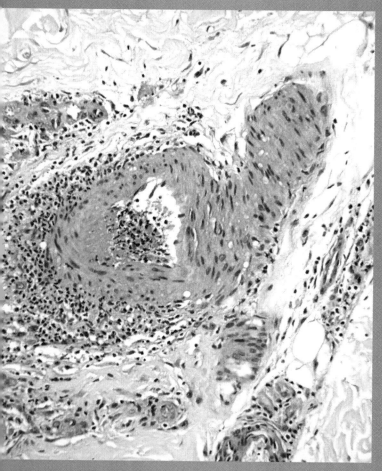

Figure 6. *Biopsy of a skin nodule.*

Laboratory: Hgb 12 g/dL, WBC 10,000/mm^3, platelet count 550,000/mm^3. ESR 65 mm/h, CRP 11 mg/dL, **Urinalysis:** normal. Chest X-ray is normal.

1. **What does the biopsy show?**

2. **What is your diagnosis?**

3. **What are the American College of Rheumatology (ACR) diagnostic criteria?**

4. **How is this condition diagnosed?**

Answers to Case 4

Biopsy result

Light micrograph of a medium sized muscular artery reveals neutrophilic infiltration with surrounding septal panniculitis. Thrombi are evident.

Diagnosis

Polyarteritis nodosa

ACR criteria

The ACR has established the following criteria for a diagnosis of PAN in a patient with vasculitis.

ACR criteria

Otherwise unexplained weight loss greater than 4 kg*

Livedo reticularis*

Testicular pain or tenderness

Myalgias (excluding that of the shoulder and hip girdle), weakness of muscles, tenderness of leg muscles, or polyneuropathy*

Mononeuropathy*

New onset diastolic blood pressure greater than 90 mmHg*

Elevated levels of serum blood urea nitrogen (>40 mg/dL or 14.3 mmol/L) or creatinine (>1.5 mg/dL or 132 μmol/L) Evidence of hepatitis B virus infection via serum antibody or antigen serology

Characteristic arteriographic abnormalities not resulting from noninflammatory disease processes

A biopsy of small or medium-sized artery containing polymorphonuclear cells*

*present in this patient

Diagnosing PAN

A sensitivity and specificity for the diagnosis of polyarteritis of 82 and 87 percent, respectively, has been found in the patient with a documented vasculitis in whom at least three of the above criteria are present.

Clinical diagnosis can be made on the basis of clinical findings and supporting laboratory features. There are no specific laboratory findings in PAN. Basic laboratory testing including complete blood counts, liver function tests, muscle enzymes, serum BUN, creatinine, and urinalysis help to determine specific organ involvement. Acute phase reactants including ESR and CRP are usually elevated, but may be normal. Blood cultures need to be done to exclude endovascular infections, such as infective endocarditis. Further laboratory testing to exclude other conditions includes: ANA, ANCA, RF, anti- CCP Ab, anti ds-DNA Ab, anti- SSA Ab, anti- SSB Ab, anti -Smith and anti-RNP autoantibodies, lupus anticoagulant (LAC), ACL Ab, and beta 2 glycoprotein-1 Ab, cryoglobulins, hepatitis B and C testing, and complements C3, C4 and CH50.

A positive ANCA argues for a diagnosis of one of the ANCA-associated vasculitides.

Due to the significant morbidity and mortality associated with the disease and complications of the therapy, biopsy confirmation of the diagnosis is sought whenever possible. Biopsy of clinically involved tissue gives the highest yield. Potential biopsy sites include cutaneous nodules, sural nerve and muscle, kidney and testicle if the latter is symptomatic. Renal biopsy may reveal medium size vessel vasculitis, however sampling error limits its value. Additionally due to the presence of renal microaneurysms, there is a potential for increased risk of bleeding. A mesenteric or renal angiogram can be performed when there is no appropriate tissue for biopsy. Characteristic angiographic findings include aneurysms, and irregular constriction in the larger vessels, with occlusion of smaller penetrating arteries.

Case 5

A 63-year-old female presents with a large swollen left knee, which developed over the last five days. She denied any trauma and had no previous history. In the past few years she has noted pain in the left knee, but this is intermittent and until last week she was able to walk a mile every day. Today she has difficulty weight bearing.

She has a long standing history of hypothyroidism and hyperparathyroidism, both of which have been well controlled. There is no history of infectious diarrhea, STDs or conjunctivitis.

Physical examination was remarkable for a warm, non erythematous, swollen, left knee, with positive ballottement of the patella consistent with an effusion. Range of motion was limited, by the swelling. Non tender Heberden's nodes of the index and middle fingers bilaterally are noted.

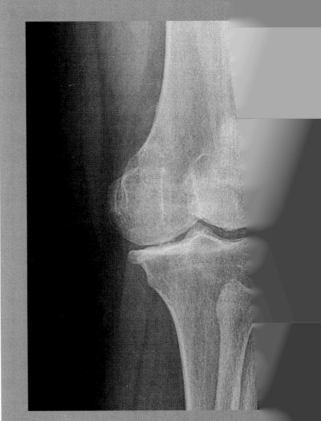

Figure 7.

X-rays illustrated in Figure 7 reveals calcification of the lateral meniscus and articular cartilage consistent with chondrocalcinosis

30 cc of cloudy fluid was aspirated from the left knee and analysis revealed 15,000 WBC with 80% neutrophils. A few intra cellular rhomboid-shaped positively birefringent crystals were seen under the polarizing microscope. Gram stain and cultures were negative

The patient responded well to an an intrarticular injection of corticosteroids.

1. **What is the most important differential diagnosis of monoarticular knee swelling, and what is the diagnosis in this case?**

2. **What are the various presentations of Calcium pyrophosphate dihydrate crystal deposition (CPPD) disease?**

3. **What are the other conditions associated with CPPD?**

Answers to Case 5

Differential diagnosis

The most important diagnosis to exclude is a septic joint which requires aspiration gram stain and culture. The most common etiologic organisms are staphylococci and streptococci, although gonococci may occur more frequently in a younger patient.

Other crystal diseases such as gout may present in this manner without necessarily having a previous episode of podagra. Reactive arthritis (conjunctivitis, diarrhea and arthritis) may cause a swollen knee, but the culture is usually sterile and there is often a history of preceding infection such as Chlamydia. Trauma also needs to be considered.

In this case the presence of weakly positively birefringent, intracellular rhomboid-shaped crystals is consistent with a diagnosis of CPPD (Calcium Pyrophosphate dihydrate Depositon disease), which was suspected based on the chondrocalcinosis seen on knee films. Calcification of the cartilage may also be seen at other sites such as the shoulder and wrist.

Presentations

In addition to the acute pseudogout presentation illustrated here, CPPD may present as pseudo-rheumatoid arthritis, pseudo-osteoarthritis, pseudo- neurotrophic arthritis and asymptomatic chondrocalcinosis.

Other conditions associated with CPPD

Strong associations occur with increasing age, previous joint surgery, OA, Gout, Trauma, Hyperparathyroidism, Hemochromatosis, Hyperphosphatasia and Hyopmagnesemia. There is a weak association with hypothyroidism.

Case 6

A 26-year-old woman with a four month history of SLE presents with painful bullous lesions on the palms and soles. At the time of diagnosis, the patient presented with a two-month history of alopecia, oral ulcers, symmetric polyarthritis, a photosensitive malar rash, and diffuse proliferative glomerulonephritis

Pertinent laboratory results:

WBC 3.0 L15% Neutrophils 56% Eos 7% Monos 12%

C3, 40 'See normal values at the end of this book'.

C4, 12 'See normal values at the end of this book'.

ANA 1:640 Speckled pattern.

Anti-ds DNA 60 'See normal values at the end of this book'.

Urinalysis: 3+ protein, WBC 0-2, RBC 3-5, RBC casts 2-5

The patient is on prednisone 40 mg daily and has been recently started on an escalating dose of mycophenolate mofetil. She now presents with painful bullous lesions on the palms and soles (Figure 8).

1. **What is your differential diagnosis?**
2. **What would be the next best step in making a diagnosis?**

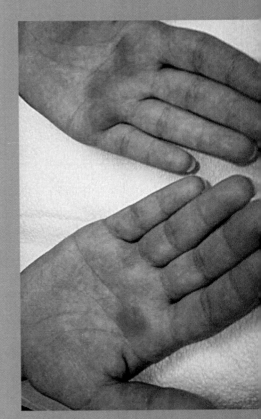

Figure 8.

Answers to Case 6

Differential diagnosis

Bullous Lupus

Bullous pemphigoid

Dermatitis herpetiformis

Epidermolysis bullosa acquisita

Steps in making diagnosis

Skin Biopsy. Subepidermal bullous lesions may be seen as a cutaneous manifestation of lupus itself, additionally lupus can rarely be associated with the other diseases listed above. In SLE, the lesions are not limited to sun exposed areas, and can be differentiated from the other conditions on the basis of histopathologic findings and immune deposits on skin biopsy, in addition to serum antibodies. Findings of continuous linear deposits of IgG at the dermal side of the epidermal basement membrane, and Ab to type VII collagen are seen in SLE and epidermolysis bullosa acquisita . Bullous pemphigoid is characterized pathologically by linear deposits of IgG and C3 at the epidermal basement membrane. Granular IgA deposits at the epidermal basement membrane characterize dermatitis herpetiformis. The presence of additional features of active Lupus suggests as is the case in this patient, that the bullous lesions are a manifestation of SLE.

Case 7

A 65-year-old man with a four year history of a progressive destructive arthritis predominantly affecting the small joints of the hands and feet, intermittent fever and scattered reddish brown papules over the hands, face, anterior chest, scalp and nose. Figure 9 Radiograph of the hands is shown.

1. **What findings are demonstrated?**

2. **What is the diagnosis?**

3. **Who is affected and what underlying conditions need to be evaluated for?**

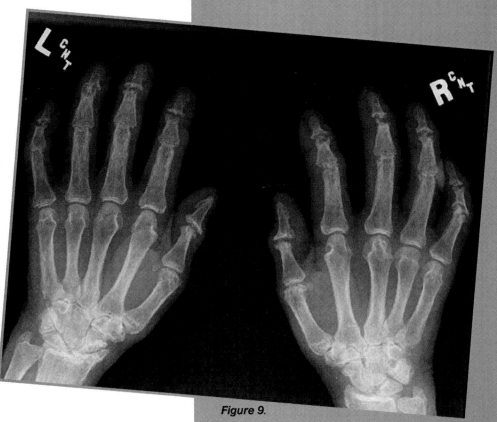

Figure 9.

Answers to Case 7

Finding

There is a symmetrical destructive arthritis with evidence for marginal erosions, and subchondral bone destruction, leading to widening of the joint spaces.

Diagnosis

Mulicentric reticulonodular histiocytosis (MRHC), also known as lipoid dermatoarthritis, is a systemic condition associated with cutaneous xanthomas, destructive arthritis, fever, and reticulohistiocytic infiltrates affecting multiple organs including the viscera (heart, gastrointestinal tract, lung), deeper soft tissues, muscle, or bone. The destructive arthritis typically involves the distal interphalangeal joints (DIP) of the hands, and feet early on. This is followed by involvement of the proximal interphalangeal (PIP) and carpal joints prior to the metacarpophalangeal joints.

The usual course is progression to arthritis mutilans. Other joints may be involved, including the shoulder, elbow, hip, and cervical spine with atlantoaxial subluxation. The articular disease can predate the skin findings. The characteristic cutaneous lesions occurring on the hands and face are reddish-brown to yellow papules and nodules. Biopsy of involved skin reveals a nodular to superficial pannicular infiltrate with multinucleated histiocytes containing abundant eosinophilic cytoplasm

Who is affected?

Women are affected three times more commonly than men, with a mean age of onset of 43 years. A paraneoplastic association is seen in up to 25% of cases including leukemia and solid tumors.

Case 8

A 50-year-old male with a chronic, progressive, destructive arthropathy of the hands and feet for 10 years had developed increased difficulty walking. He had difficulty opening jars, and walked with a limp. He was stiff for over one hour every morning. Examination revealed multiple flexion deformities of PIPs, and bony deformities and fusion of DIPs. He had difficulty making a full fist. He had synovitis and marked decreased ROM of his wrists.
A patch of a scaly rash at the base of the scalp, as well as in his umbilicus was noted. Nail pitting was observed.

Dactylitis (diffuse swelling of the digit) of several toes was seen.

Labs show ESR 35 mm/hr
X rays of hands Figure 10 reveal erosions of MCPs PIPs and DIPs with subluxations of MCPs and classic pencil in cup deformity at the right 2nd and 3rd DIPs.

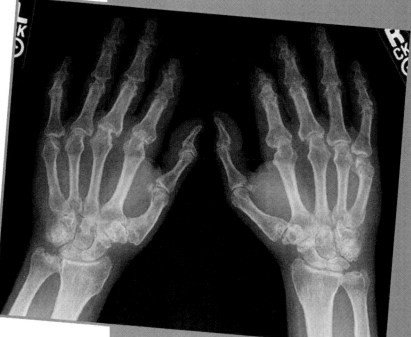

Figure 10.

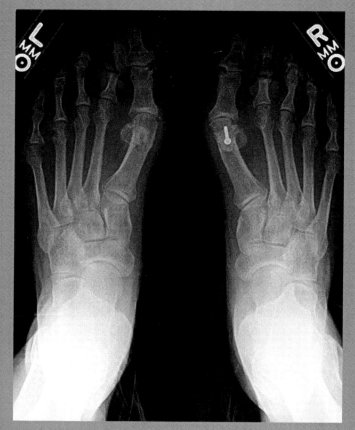

Figure 11.

X-rays of feet in Figure 11, show pencil in cup deformity of left 3rd toe and severe erosive changes of MTPS and IP Joints.

1. **How does Psoriatic Arthritis (PSA) differ from Rheumatoid Arthritis (RA)?**

2. **What are the characteristic clinical findings that help characterize Psoriatic Arthritis?**

3. **What is the differential diagnosis of Psoriatic arthritis?**

4. **What are some of the treatment options for this patient?**

Answers to Case 8

Differences between PSA and RA

Psoriatic arthritis may mimic rheumatoid arthritis but is usually more asymmetric in presentation. There may be one or two joints involved, or it may be very aggressive and progress to "arthritis mutilans" with significant joint damage, as in this case. Sacroliliitis may be a feature of psoriatic arthritis but is not usually a feature of rheumatoid arthritis. Psoriatic arthritis may present with inflammatory and erosive changes at the DIPs, unlike rheumatoid arthritis where the damage typically spares the DIPs. Other features of this inflammatory arthritis may include sausage digits (dactylitis) – diffuse swelling of the digit.

Clinical Findings

Nail pitting and onycholysis are also common. In addition to the inflammatory arthritis, patches of psoriasis commonly occur on the extensor surfaces of elbows and knees and may be widespread. Occult areas of psoriasis should be looked for in the umbilicus, scalp, intergluteal cleft and perineum.

Differential Diagnosis

Rheumatoid Arthritis, Reactive arthritis and Viral Arthritis, such as Parvovirus.

Treatment Options

This patient appears to have an aggressive course of disease with erosions.

A disease modifying agent such as methotrexate would be initial therapy with addition of a biologic, such as a TNF-inhibitor, if rapid resolution of swelling and tenderness does not occur. The goal of therapy is to prevent further joint damage and restore function. Non steroidal Anti-Inflammatory medications (NSAIDs) and physical therapy may also help improve range of motion while waiting for the disease modifying drugs (DMARDs) to take effect.

Case 9

A 54-year-old woman has a ten-year history of severe joint pains and swelling, which have resulted in difficulty using her hands. She admits to two hours of morning stiffness and her hands are now twisted. Recently, she has difficulty walking because of pain and swelling in her feet. She states that it feels like she is "walking on marbles". Her knees swell intermittently and she has to use a cane on a daily basis. In addition, she has noted lumps on her elbows. Driving has become more difficult due to limited rotation of her neck.

Examination of her hands reveals bilateral ulnar deviation with subluxations that are not reducible. Firm nodules are seen on the elbows,

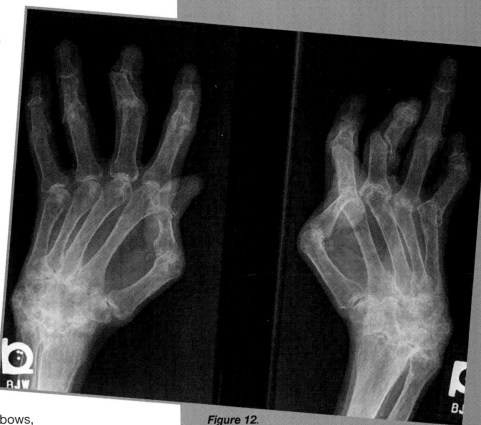

Figure 12.

and the toes show cock up deformities. Cloudy fluid was aspirated from her swollen right knee.

Lab revealed an elevated erythrocyte sedimentation rate (ESR) of 60 mm/hr. A stable normochromic, normocytic anemia is noted. Liver and renal function are both normal. Synovial fluid showed 25,000 WBCs with no crystals identified.

X-rays are illustrated In Figure 12. They show severe joint deformities with subluxations of the MCPS and fusion of the carpal bones and wrists. Erosions of the PIPS are also noted. These findings are long standing and most consistent with arthritis mutilans due to rheumatoid arthritis.

1. **What are some of the prognostic factors suggesting aggressive disease?**

2. **Which disease modifying therapies would be recommended for this patient?**

3. **If the patient needs surgery, what suggestions should be made with regard to therapies and what are their principal side-effects?**

Answers to Case 9

Prognostic factors

Poor prognostic factors include: highly positive rheumatoid factor, positive anti-CCP antibody, multi-joint involvement and extra-articular manifestations such as rheumatoid nodules and vasculitis.

Therapies

The above patient has multiple poor prognostic factors including a highly positive rheumatoid factor and positive anti-CCP antibody, and needs aggressive treatment with disease modifying agents. She has already been taking maximal methotrexate therapy, 20 mg once a week with daily folic acid, along with etanercept 50 mg once a week. She still has active inflammation as well as chronic deformities and the goal of therapy would be to prevent further joint damage. Options would include changing to a different anti-TNF agent such as infliximab or adalimumab. Biologic therapies such as abatacept (a T cell co-stimulator inhibitor) or rituximab (a CD 20 B cell inhibitor) are other options. Glucocorticoids are avoided if possible in the treatment of RA because of the myriad of side-effects and difficulty tapering once started. However, for severe exacerbations they may be used temporarily at low doses as bridge therapy until effective adjustment of additional DMARDs.

Therapies prior to surgery and principal side-effects

Biologic therapies should be held at least one to four weeks before surgery to decrease the risk of infection, and only restarted one to four weeks afterwards depending on which biologic is being used. Infections are the commonest side-effects of these medications.

Other side-effects related to TNF-alpha inhibitors include tuberculosis re-activation, drug induced lupus and demyelinating disorders. There may be an increased risk of skin cancers and lymphoma, although patients with longstanding RA are also at risk of the latter.

Patients with an underlying diagnosis of COPD are at increased risk of exacerbation and possibly lung cancer thus abatacept therapy should be used with caution in this population. Infusion reactions may be an issue with rituximab and pre-medication with steroids is needed. Non steroidal anti-inflammatory medications (NSAIDs) should be discontinued one week prior to surgery to decrease the risk of bleeding. Any patient taking cortico-steroids would require stress dose. Flexion and extension neck films are important to look for atlanto-axial subluxation, whose presence is to be communicated urgently to the anesthesiologist. Avoidance of hyper-extension of the neck during anesthesia is imperative to prevent neurologic sequelae.

Case 10

A 32-year-old male has five years of low back pain, worse on the right side, particularly in the morning when he gets up. He is stiff for about one hour. He improves as the day progresses, and feels better with activity. He has a history of a swollen left knee in the past and synovial fluid aspiration, but the results are unknown. In addition, he intermittently has red eyes. He had an episode of "food poisoning" about one month prior to the onset of his joint pains. He remembers it well as he was admitted to the hospital for several days for intravenous fluid replacement. He was told he was dehydrated at the time, from his severe diarrhea. He does not remember if he was treated with antibiotics. There is a family history of low back pain. He remembers his paternal uncle walking with a stooped appearance.

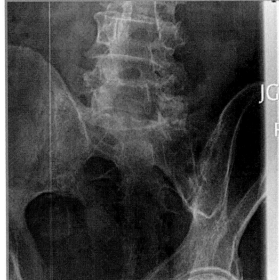

Figure 13.

On examination he has tenderness over the right sacroiliac joint with limited flexion of the spine. He has warmth and swelling of the right knee.

The X-ray is an oblique view of the right sacroiliac joint demonstrating erosions and sclerosis of the joint.

1. **What is the differential diagnosis?**

2. **What treatment is suggested?**

3. **What organisms can predispose to reactive arthritis?**

Answers to Case 10

Differential Diagnosis

The patient most likely has a reactive arthritis manifested by the knee swelling after the episode of dysentery. The sacroiliitis, seen on the X-ray and positive family history of similar problems suggests that this disease is likely HLA-B27 related. Other possibilities include psoriatic arthritis, which would be unusual without rash, (although rarely the inflammatory arthritis and/or sacroiliitis can precede the skin manifestations of psoriasis).

Other manifestations of reactive arthritis may include enthesopathy, mucosal and ocular lesions as well as cutaneous manifestations, such as keratoderma blennorrhagicum. Enthesopathy refers to inflammation at the point of attachment of tendons and ligaments to bone, and is a feature of all of the spondyloarthropathies. Reactive arthritis is one of the spondyloarthropathies.

Other spondyloarthropathies include ankylosing spondylitis, the spondyloarthritis of inflammatory bowel diseases – ulcerative colitis and Crohn's disease, and psoriatic spondyloarthritis.

Treatment

Anti-inflammatory medications and physical therapy are usually recommended as first line therapy. However, with recurrent episodes of swollen knees and the documented sacroiliitis it is likely that he will need a disease modifying agent.

Organisms that predispose to ReA (reactive arthritis)

Shigella, Salmonella, Yersinia, Campylobacter and Chlamydia have all been implicated in the development of reactive arthritis. Episodes of reactive arthritis may occur one to four weeks after an episode of urethritis or dysentery. Approximately fifty percent of patients with reactive arthritis experience spontaneous resolution of symptoms within six months. A form of chronic reactive arthritis may develop in some patients characterized by a relapsing disease course.

Case 11

A 56-year-old African American woman with a history of systemic lupus erythematosus and discoid lupus (primarily involving the face and scalp) presented with tender and painful subcutaneous nodules on her face, forehead and scalp. She states that the lesions started approximately six months prior. The lesions had started to ulcerate two months prior to the visit. She had previously been on hydroxychloroquine for her systemic lupus erythematosus and discoid lupus with good control, but states that she ran out of her medication about 12 months ago. Her only current medication is over the counter naproxen for the pain.

Cutaneous Exam:

Her skin exam reveals multiple ulcerative lesions in various stages of healing on her face, forehead, and scalp (See figures 14 and 15). The lesions have caused disfigurement. She has no skin lesions elsewhere on her body.

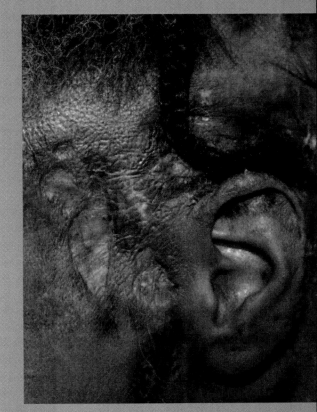

Figure 14.

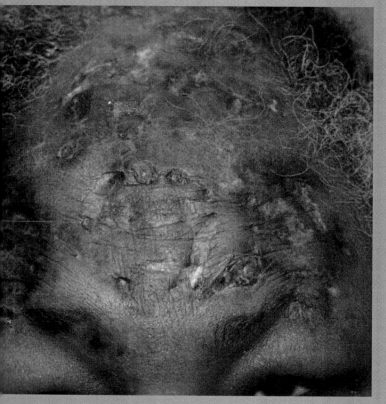

Figure 15.

Pathology:

A skin biopsy was performed. Histologic features included epidermal atrophy, hydropic degeneration of the basal cell layer of the epidermis, and lymphocytic inflammation that extends into the subcutaneous fat.

1. **What is the diagnosis and what are the clinical features?**

2. **What is a major clinical feature differentiating lupus panniculitis from another form of panniculitis, erythema nodosum?**

3. **Do the majority of patients with systemic lupus develop lupus panniculitis?**

4. **How do you treat lupus panniculitis?**

Diagnosis and clinical features

Lupus panniculitis, or lupus profundus, primarily affects subcutaneous fat. In nearly all cases there are deep, erythematous plaques and nodules, and some ulcers, which usually involve the proximal extremities, trunk, breasts, buttocks, face, and scalp. Patients with lupus panniculitis often have a history of concomitant discoid lupus. Lesions may be tender and painful and frequently heal with atrophy and scars.

Major clinical feature differentiating lupus panniculitis

Erythema nodosum most frequently occurs on the lower extremities and the inflammation involves the septa of the fat lobule, whereas lupus panniculitis only rarely occurs on the lower extremities and the inflammation involves the fat lobule. There is also no strong association with erythema nodosum and systemic lupus or discoid lupus. The main differential diagnosis is subcutaneous panniculitis-like T-cell lymphoma (SPTCL).

Systemic lupus

No. Lupus panniculitis is a rare manifestation of systemic lupus. Approximately 1-3% of patients with systemic lupus develop lupus panniculitis.

Treatment

Lupus panniculitis typically responds to the same treatments utilized for other forms of cutaneous lupus. The mainstay of treatment is often anti-malarials, such as hydroxychloroquine. Quinacrine is sometimes used as well. Corticosteroids are often helpful to control the acute symptoms, but intralesional corticosteroids should be avoided because they are generally ineffective and may exacerbate the atrophic healing process. For refractory cases dapsone, azathioprine, and thalidomide have been utilized.

Case 12

This patient is a 25-year-old African American female who, at the age of 17, originally developed pain, stiffness, and swelling in her right knee. Over the ensuing year she developed similar symptoms in her left knee followed by a more symmetrical polyarthritis that also involved her hands, wrists, and elbows. She was diagnosed with rheumatoid arthritis (RA) at that point and started on Methotrexate (MTX).

She discontinued her MTX (15mg weekly) approximately six months later secondary to a perceived lack of efficacy. At the age of 21, she was started on etanercept. She had an initial response to this medication; however, it was discontinued one and a half years later because the beneficial response had waned. She experienced no side-effect to etanercept.

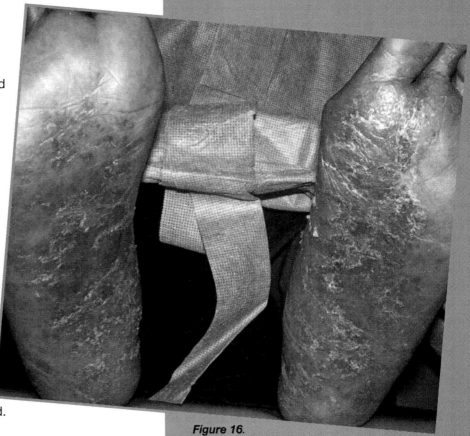

Figure 16.

At the age of 23 (one and half years prior to presentation), the patient was started on infliximab monotherapy for her RA. Her arthritis improved on infliximab, and she remains on this drug.

For the past seven months the patient began to experience erythematous pustules on the plantar surface of her left foot, followed by similar lesions on the right foot a few weeks later. Over the next four months the skin lesions worsened. She began to experience diffuse pustular lesions with scaling and excoriation; however, the lesions remained confined to the plantar surface of her feet (see Figure 16).

1. **What is your diagnosis?**

2. **What is the most typical presentation of these uncommon, yet well-documented, skin lesions that occur as a side-effect to the TNF-alpha antagonists?**

3. **What other drugs are known to both treat and cause de novo cases of the same disease?**

4. **What is the etiology of this cutaneous side-effect of anti-TNF therapy?**

Answers to Case 12

Diagnosis

This patient has an uncommon plantar **pustular psoriasiform eruption** that can occur in patients undergoing treatment with tumor necrosis factor (TNF)-α antagonists. TNF-α antagonists have revolutionized the treatment of many chronic inflammatory conditions, including rheumatoid arthritis (RA), psoriatic arthritis, and Crohn's disease. Despite such successes, use of these agents has been associated with development of several cutaneous conditions, including skin infections, skin tumors, and vasculitis. Much recent interest has been directed to the relatively uncommon pustular psoriasiform eruptions that can occur in patients undergoing treatment with TNF-α antagonists. Given that these same agents have proved to be efficacious in the treatment of psoriasis, it is unclear how they can cause de novo "psoriasis". In the literature, this puzzling side-effect has been described as a "paradox" and a "conundrum".

While the true etiology underlying these lesions remains elusive, several points of clarity do exist. First, this side-effect has been described in patients undergoing treatment with each of the three TNF-α antagonists (etanercept, infliximab, and adalimumab), indicating that it represents a class effect. Second, a disproportionately high percentage (52%; 31/60) of such psoriasiform lesions involves the palms and/or soles; however, palmoplantar pustular psoriasis represents only 1.7% of all cases of psoriasis. Finally, the lesions persist while patients are undergoing treatment with a TNF-alpha antagonist but resolve when the drug is discontinued.

Typical presentation

The most typical presentation is of pustular scaley skin lesions that resemble palmoplantar pustular psoriasis. The majority of these cases occur on the palms and soles. Although these skin lesions are typically referred to as "psoriasis" or "psoriasiform", the palmoplantar location that occurs in the setting of anti-TNF-alpha therapy contrasts with idopathic psoriasis in that palmar and pustular involvement is unusual in idiopathic psoriasis, occurring in less than 2% of cases.

Other drugs known to both treat and cause de novo cases of the same disease

None.

Etiology of side-effects of anti-TNF therapy

The true etiology remains elusive. As stated, one theory is that these are de novo cases of psoriasis, yet there is no proven pathophysiology to support it. Other theories suggest that this might represent an exacerbation of a latent infection, such as with *Chlamydia trachomatis*. It is well known that chlamydial infections can exist in a persistent, yet aberrant state. Chlamydial replication is inversely related to TNF levels in vitro. Keratoderma blennorrhagicum (KB) is a cutaneous palmoplantar pustular rash that occurs in chlamydia-induced reactive arthritis. Interestingly, KB is both grossly and histologically identical to pustular psoriasis.

Case 13

A 24-year-old woman with a history of Crohn's Disease of the terminal ileum with a rectovaginal fistula refractory to treatment with corticosteroids and mesalamine presented to the Emergency Department nine days after her first dose of infliximab. On the evening she received her first dose of infliximab (5mg /Kg), she experienced arthralgias in her wrists, MCP's, PIP's and ankles with paresthesias of the hands and feet.

These symptoms resolved the following morning. Nine days later she presented with a painful rash involving the arms and legs. The rash was accompanied with pain and swelling of the hands and feet as well as generalized myalgias. She complained of a subjective fever and mild abdominal pain. She had not recently traveled and had no sick contacts. She had not recently started any other new medications.

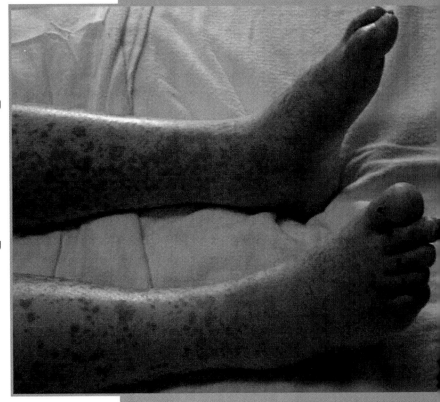

Figure 17.

Cutaneous Exam

Skin exam revealed diffuse palpable purpura over the legs (Figure 17) and posterior aspects of the arms. The lesions were non-blanchable.

Labs:

Her erythrocyte sedimentation rate was 48mm/h and the CRP was 5mg/dL (normal <0.5mg/dL). Antinuclear antibodies, ds-DNA and histone antibody were negative.

Pathology:

Fragmented neutrophilic infiltrate (nuclear debris) of small vessel walls.

1. What is your diagnosis and what is the etiology?

2. Is it unusual to develop a reaction to a drug nine days after exposure?

3. Besides the skin, what other organ systems can be involved with Hypersensitivity Vasculitis (HV)?

4. Could this patient's HV have been caused by something other than infliximab?

Answers to Case 13

Diagnosis and Etiology

The diagnosis in this patient is hypersensitivity vasculitis (HV), which involves inflammation of the small vessels, mostly post capillary venules of the skin and other organs. It is similar in etiology to other immune complex disorders due to circulating antigen-antibody immune complexes with vessel wall deposition. This results in ischemia of tissues supplied by the inflamed vessel. HV can manifest as multisystem disease or may involve only the skin. It presents as palpable purpura and is associated with fever, malaise, myalgias, and arthralgias.

It is important to note that leukocytoclastic vasculitis (LCV) is a histopathologic finding, not a clinical diagnosis. It involves deposition of immune complexes in vessel walls, ultimately leading to cellular infiltrates, cytokine release, and vessel damage. HV most commonly occurs secondary to drug reactions. Drugs that have been implicated include the penicilins, sulfonamides, thiazides, NSAIDs, phenytoin, and more recently the TNF-alpha antagonists.

Unusual Reaction

While this is somewhat unusual, it must be kept in mind that infliximab has a long half-life (~9 days). Her symptoms that began nine days after exposure to the drug were consistent with a type III Gell and Coombs immune response involving immune complex formation with subsequent complement activation and release of lysosomal enzymes. The antibody formation can take several days which might explain why this patient developed the reaction nine days after the initial exposure.

Extracutaneous Involvement

The internal organs most commonly affected include the gastrointestinal tract and kidneys. Additionally the joints may be affected.

Possible Other Causes

Crohn's disease itself has rarely been associated with HV. However, more common cutaneous manifestations of Crohn's disease include pyoderma gangrenosum and erythema nodosum. Considering the rash developed nine days after her exposure to infliximab, however, makes infliximab the most likely cause. After successful treatment, the patient did not receive any more infliximab and has not developed another case of HV for four years.

Case 14

A 48-year-old Caucasian female presented four years prior with a painful, injected right eye that was associated with photophobia. She was diagnosed with scleritis and treated with topical corticosteroids. The scleritis improved. Later, her scleritis returned, was bilateral, and became refractory to treatment with both topical and systemic corticosteroids. Anytime she was on prednisone doses of less than 30mg daily, her scleritis would flare. Her symptoms improved after methotrexate was added and she was able to completely wean off corticosteroids. She remained asymptomatic for approximately six months until she developed headaches and right upper extremity parasthesias. She denied fevers/chills, diplopia, nausea or vomiting.

Neurologic Examination:

On presentation to clinic, she was alert and oriented. Her speech was clear. Her cranial nerves were intact. Motor exam revealed 5/5 strength in the upper and lower extremities. Sensation was normal to light touch and pinprick throughout. Reflexes were brisk and symmetric. Her gait was normal.

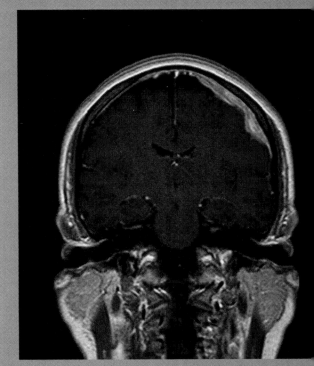

Figure 18.

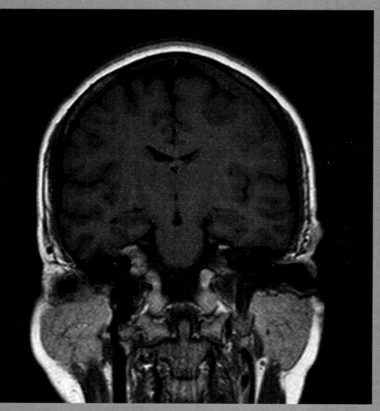

Figure 19.

Laboratory studies:

Blood work showed a white blood cell count of 12.1 k/ul with granulocyte predominance. Serum electrolytes were within normal limits. Her coagulation profile was unremarkable. Erythrocyte sedimentation rate (ESR) was 125. CSF cultures and tests for acid fast bacilli were negative by polymerase chain reaction method. C-ANCA and P-ANCA were negative.

Imaging:

A chest X-ray did not reveal any pulmonary infiltrates or hilar adenopathy. MRI of the brain with and without gadolinium enhancement (Figure 18 and 19) revealed a homogeneously- enhancing extra-axial mass that tracked along the left fronto-parietal regions of the convexity. There was diffuse thickening of the dura at multiple locations, with minimal mass effect on the left parietal lobe. No significant, associated hyperostosis was identified. There were no mid-line nasal lesions.

Neuropathology:

Initial intra-operative consultation by pathology on the intra-operative specimen showed fibrous tissue with necrotizing inflammatory lesions. No tumor was identified. Examination of routinely processed, paraffin embedded tissue revealed necrotizing granulomatous inflammation with acute and chronic inflammatory components. The granulomas were poorly defined with extensive necrosis, and only scattered multinucleated giant cells. Special stains for bacteria including spirochetes and Tropheryma whipplei, acid fast bacilli as well as stains for fungi yielded no positive results.

1. **What is the differential diagnosis and features which support the final diagnosis?**

2. **What is the final diagnosis?**

3. **Are all patients with Wegener's granulomatosis positive for C-ANCA?**

4. **Is it unusual for Wegener's patients to develop eye symptoms?**

5. **How often does Wegener's include neurologic involvement?**

Answers to Case 14

Differential Diagnosis

Diagnostic considerations in this patient include infectious meningitis, meningeal thickening associated with neurosyphilis, primary or secondary dural tumors and rheumatologic diseases - particularly neurosarcoidosis, and hypertrophic cranial pachymeningitis with associated vasculitis. CNS infections usually cause meningitis with diffuse dural enhancement. Abscesses manifest as intraparenchymal, space occupying lesions. The clinical picture associated with infections with or without abscess formation is often quite distinct and is inconsistent with this patient's presentation (e.g., absence of fevers, mental status changes, and focal deficits). CNS gumma lesions, with focal dural thickening, have been described with neurosyphilis. Laboratory work-up and neurologic exam, however, argue against an infectious etiology.

As stated, primary and metastatic tumors are also on the differential diagnosis. Meningioma, metastatic tumor and rarely lymphoma may manifest in a predominantly dural, plaque-like enhancement. Clinical features as well as associated imaging studies will often help resolve these differential diagnoses. Meningiomas cause focal thickening and enhancement of the dura.

The MRI characteristics in this case would favor an en plaque meningioma. However the absence of hyperostosis on imaging studies would speak against it. Primary CNS lymphoma (PCNL) usually presents as solitary or multiple supratentorial lesions. Imaging characteristics, however, do not favor this, as isolated meningeal enhancement is rare in PCNSL. Dural-based metastatic lesions can have similar MRI characteristics. With the lack of known or suspected primary tumor, however, the probability of this being a metastatic lesion is unlikely.

Neurosarcoidosis often presents with an enhancing pattern tracking along the sulcal and gyral contours of the brain, reflecting pial abnormality, although extra-axial dural granulomata can also be seen and are clinically difficult to distinguish. CNS involvement in systemic vasculitic disorders is well-established. Examples of these entities include Wegener granulomatosis and primary angiitis of the CNS. The patient had no clinical evidence of pulmonary or renal disease, but cases of isolated CNS vasculitis have been reported in several types of rheumatic conditions.

Final Neuropathologic Diagnosis:

Dural granulomatosis with vasculitis most consistent with intracranial Wegener's granulomatosis.

Patients with C-ANCA

No. C-ANCA is about 80% sensitive and 95% specific for this disease in patients with renal or pulmonary involvement. As many as 40% of patients with extra-renal or extra-pulmonary disease are C-ANCA negative.

Eye Symptoms

No. The eye is the fourth most commonly involved organ system (after renal, upper respiratory, and lower respiratory involvement), often in the form of scleritis.

Neurologic Involvement

Neurologic involvement has been reported as high as 33.6% of cases.

Case 15

The patient is a 41-year-old Caucasian female who presented two and a half years earlier with headaches, fatigue, diffuse myalgias and arthralgias, galactorrhea, and short term memory problems. Her initial physical examination revealed an obese female (BMI 38.97) with normal vital signs. She had no proptosis, extraocular muscle dysfunction, thyromegaly, or thyroid masses.

The remainder of her physical examination was unremarkable. She was found to have an elevated serum prolactin level of 52ng/ml (normal 3-30ng/ml for a premenopausal non-pregnant female) and a normal thyroid stimulating hormone (TSH). Her estradiol level was <1.0 pg/ml (normal 10-400pg/ml) and her follicle stimulating hormone (FSH) was 1.8mIU/ml (normal 1.5-18.8mIU/ml). An MRI of the brain revealed the pituitary to be 13mm x 9mm in size (upper limit of normal) without any definite mass. The patient was felt to have a pituitary microprolactinoma complicated by anovulation and galactorrhea.

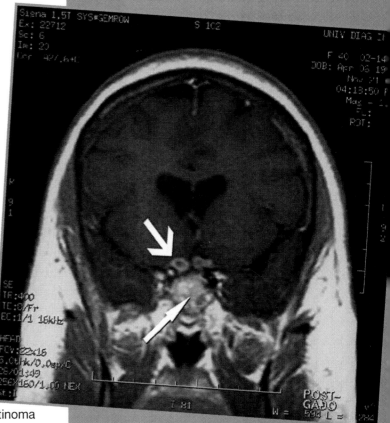

Figure 20.

After failing two months of medical therapy with bromocriptine, a repeat MRI of the brain was obtained that showed an enlarged pituitary compared to the previous study (16mm x 12mm). It was also felt to be a true right-sided pituitary mass now present.

A transsphenoidal resection of the pituitary mass was planned. During the surgery, there was no evidence of any pituitary adenoma after opening the dura. The pituitary gland was dissected 3mm deep and a firm mass was recognized. Multiple biopsies were obtained and the frozen sections revealed multiple non-caseating granulomas.

The final pathology showed significant non-necrotizing granulomata and mononuclear inflammatory infiltrates. There were also multinucleated cells without evidence of Touton giant cells. The PAS and AFB stains were negative. Focal synaptophysin immunoreactivity confirmed the presence of pituitary tissue. A diagnosis of neurosarcoidosis was made.

The patient was initially treated with prednisone 60mg daily. Her headaches, fatigue, and arthralgias improved. After two months, the patient was unable to decrease her prednisone dose below 20mg daily because of increased headaches, arthralgias, and weakness. Oral methotrexate (7.5mg weekly) was added and the dose was titrated to 15mg weekly over three months. The patient's symptoms worsened and she began to experience visual loss with left homonymous superior quadrant hemianopsia. She was found to have optic neuropathy, right eye more so than the left.

A repeat MRI of the brain now showed multiple areas of prominent enhancement including the suprasellar and parasellar region [thin arrow; Figure 20], intracranial pre-chiasmal optic nerves (right > left) and optic chiasm [thick arrow Figure 20], intracranial fifth nerve bilaterally, cavernous sinus, and infundibulum. The patient's methotrexate was discontinued and she was started on cyclophosphamide 1000mg intravenously monthly.

1. **When patients with known sarcoidosis present with new neurological symptoms, is the diagnosis neurosarcoidosis?**

2. **Does neurosarcoidosis always require treatment?**

3. **Can a diagnosis of neurosarcoid be made by histologic evaluation?**

Answers to Case 15

Sarcoidosis and neurological symptoms

While it is true that many patients with known sarcoidosis who present with new neurological symptoms will have neurosarcoidosis, unrelated diseases (especially infection) need to be ruled out. Neurosarcoidosis occurs in a minority of patients with sarcoidosis, about 5%. However, it is serious and portends a poor prognosis. We present a patient with neurosarcoidosis that was refractory to treatment with corticosteroids and methotrexate.

About two thirds of patients with neurosarcoidosis have a self-limited monophasic illness, and the rest have a chronic remitting relapsing course. With treatment, death from neurologic disease is unusual. When sarcoidosis involves the nervous system, there is a wide range of symptoms, but they usually include cranial nerve involvement, peripheral nerve involvement, and/or central nervous system involvement. The cranial nerves most often involved include the facial nerve, optic nerves, and auditory nerves. The peripheral nerve involvement can be of sensory or motor nerves. CNS involvement includes the hypothalamus/pituitary, cerebral cortex, cerebellum, and spinal cord (rarely).

Treatment for neurosarcoidosis

Yes, neurosarcoidosis has a poor prognosis if untreated. The initial mainstays of treatment include corticosteroids plus either azathioprine, methotrexate, or cyclophosphamide. If the disease is refractory to these conventional treatments, emerging data suggest infliximab is effective, but it is imperative that systemic *Mycobacterium tuberculosis* (and other infections that are obvious) be ruled out prior to treatment.

Histologic evaluation of neurosarcoid

Pathology is an important contributor to the diagnosis, but it must be coupled with clinical symptoms, advanced imaging (such as MRI), and the investigation of an underlying infectious process in order to make the proper diagnosis. Typical histologic features include lymphocytes and mononuclear phagocytes surrounding a non-caseating epithelioid cell granuloma. The diagnosis is made by the typical clinical signs and symptoms as well as histologic analysis. MRI is the imaging modality of choice because of the superior images obtained that include the characteristic findings such as those seen in this patient. These include periventricular high intensity lesions on T2 weighted images, multiple supratentorial or infratentorial lesions, solitary intra or extra-axial masses, leptomeningeal enhancement, optic nerve enhancement, or spinal cord intramedullary masses.

Case 16

A 25-year-old patient is referred to you with a painful left thigh of six weeks' standing. She brings in X-rays from another facility, that are remarkable for an Erlenmeyer flask deformity of the distal left femur (Figure 21).

Your examination reveals hepatosplenomegaly, and there is anemia and thrombocytopenia on her complete blood count. You send a blood sample to your reference laboratory for further testing, that comes back showing a deficiency of acid beta-glucosidase in nucleated cells. You make a diagnosis of Gaucher's disease and refer the patient to an endocrinologist for treatment.

1. **What is the underlying genetic disorder, and describe the clinical manifestations of Gaucher's disease?**

2. **What are the radiographic findings seen in Gaucher's disease?**

3. **How is a diagnosis made?**

4. **What is the treatment?**

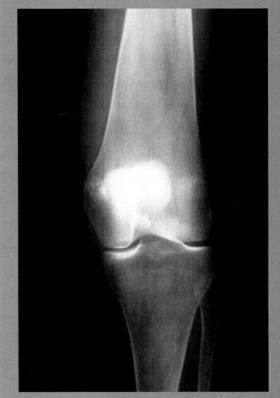

Figure 21.

Answers to Case 16

Underlying genetic disorder

Gaucher's disease is the most common lysosomal storage disease. There is deposition of glucosylceramide (also called glucocerebroside), within the lysosomes of cells. Gaucher's disease is caused by various mutations leading to a deficiency of the enzyme acid beta-glucosidase. It is inherited as an autosomal recessive disorder and has a bimodal age predilection: younger than 15 years and about 25 years of age. Gaucher's disease is seen in 1 in 1000 Ashkenazi Jews and in less than 1 in 100,000 in other ethnicities.

A N370S mutation (asparagine to serine mutation on chromosome 1) is the most common mutation in Ashkenazi Jewish and Spanish patients. Bone involvement in Gaucher's disease is of major impact. Osteopenia and osteonecrosis are problems encountered in Gaucher's disease. Osteonecrosis of the femoral head with femoral neck fracture is common. Osteopenia or osteonecrosis of the vertebrae can cause compression fractures. Bone crises with acute punctuations of severe, localized pain from bone infarction occur. There is also chronic bone pain.

Hepatoslenomegaly can be mild or massive. Splenic infarction can mimic an acute abdomen. Anemia and thrombocytopenia are seen.

Radiographic findings

X-rays may demonstrate the characteristic Erlenmeyer flask deformity of the distal femur caused by abnormal modeling of the metaphysis, as seen in our patient.

Diagnosis

Decreased acid beta- glucosidase activity (0 – 20% of normal) in nucleated cells is diagnostic. Gaucher cells in the bone marrow are characteristic of this disease, and have the appearance of wrinkled paper from deposition of glucosylceramide.

Treatment

Enzyme replacement for severe cases is effective and safe. Transfusions are called for in some cases. Joint replacement surgery can be beneficial, and bisphosphonates improve bone density.

Case 17

A 50-year-old, white male college professor notices polyuria and weight loss of 12 pounds over the preceding three months. He also describes darkening of his skin without unusual sun exposure. There is pain in his metacarpophalangeal (MCP) joints for six weeks. The pt complains that his sexual drive is not what it used to be, and he does not awaken mornings with an erection anymore. The patient is adopted. On examination, the skin is bronzed. The second and third MCP joints are enlarged, hard, and tender to palpitation.

X-rays (Figure 22) show peculiar hook-shaped osteophytes of the second and third MCPs bilaterally.

Laboratory results were normal except for a transferrin saturation of 65%.

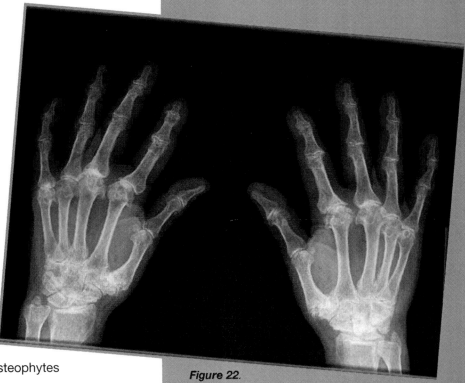

Figure 22.

1. What is the diagnosis and genetic basis of the underlying disorder?

2. What are the musculoskeletal features of this disease?

3. What are the other clinical features of this disease?

4. How is the diagnosis made?

5. What is the treatment?

Answers to Case 17

Diagnosis

This patient has hereditary hemochromatosis (HH), which is an inborn error of iron metabolism, most often caused by homozygosity for the C282Y (cysteine replaced by tyrosine at amino acid position 282) mutation of the HFE-gene on chromosome 6. Another mutation responsible for HH is the C282Y/H63D compound heterozygous state, an H63D mutation resulting in a substitution of histidine to aspartic acid at amino acid position 63. HH is inherited as an autosomal recessive trait, and it is one of the most common heritable metabolic diseases, with a prevalence of about 3 per 1000 population.

This genetic abnormality results in increased intestinal iron absorption, and iron overload. Men are affected 10 times more commonly than females. The decreased prevalence in females is most likely related to the protective effect of the iron loss with menstruation. There is a predilection for people of northern European descent. The age of onset of symptoms is most often older than 40 years.

Musculoskeletal features

The joints are involved in 50% of patients with hereditary hemochromatosis and may be the presenting symptom. Osteoarthritis and calcium pyrophosphate dihydrate crystal deposition disease (pseudogout) are part of the arthropathy of HH. The pathogenesis is not clear, and iron deposition is not the entire explanation for the arthropathy of HH. Clinically, a usual presentation is pain on flexing the second and third metacarpophalangeal joints. Joint inflammation is usually mild. The large joints, such as the hips, knees and shoulders, can be involved and arthroplasty might be necessary.

On X-ray, there are characteristic hook-shaped osteophytes of the second and third MCP joints (seen on this patient's X-ray). In fact, osteoarthritis of the second and third MCP joints, in the absence of a history of trauma, indicates hemochromatosis -related arthropathy until proven otherwise. In addition, there is chondrocalcinosis, calcification of the hyaline and fibro cartilage on X-ray. Osteoporosis occurs in 25 – 50% of patients with HH. The pathogenesis of this abnormality is also unclear. Osteoporosis can occur in HH in the absence of hypogonadism.

Clinical Features

Iron accumulates in the parenchyma of various organs, with consequent organ dysfunction. Cardinal extraarticular clinical manifestations of hemochromatosis include:

1. **Diabetes mellitus from iron deposition in the pancreas**

2. **Cirrhosis of the liver from iron deposition in the liver, with ensuing hepatocellular carcinoma in some cases**

3. **Hypogonadotropic hypogonadism from iron deposition in the FSH/LH secreting cells of the pituitary gland**

4. **A restrictive cardiomyopathy from iron deposition in the heart**

5. **Bronzing of the skin from an increase in melanin, leading to the so-called "bronzed diabetic" patient.**

Making the diagnosis

Hemochromatosis is diagnosed first by assaying fasting transferrin saturation, then assaying for the HFE-gene C282Y mutation mentioned above if the blood transferrin saturation is greater than 60% in men, 50% in women.

Less commonly, other gene mutations are responsible for HH, such as the C282Y/H63D compound heterozygote, also discussed above. Liver biopsy can indicate the degree of iron deposition and cirrhosis, and magnetic resonance imaging (MRI) detects intracellular iron deposits, which reduce the parenchymal T2 signal. The liver signal is reduced to less than that of adjacent muscle. With evolving techniques, MRI will have a greater role in the diagnosis of hemochromatosis.

Treatment

Hemochromatosis is treated by phlebotomy or chelation of iron with deferoxamine if phlebotomy is not feasible. Removal of iron might improve the fatigue, right upper quadrant discomfort, hepatomegaly, and elevated liver enzymes, and cardiac function might also improve if treatment is begun before cardiomyopathy occurs.

Diabetes mellitus is often easier to control after iron removal. However, the arthropathy, established cirrhosis, and testicular atrophy generally do not improve. The arthropathy is treated symptomatically with analgesics and NSAIDs. Arthroplasty might be necessary, but an increased incidence of loosening of the joint prosthesis is seen in HH. In the pipeline is the therapeutic use of hepcidin to prevent iron overload.

Case 18

A 70-year-old Caucasian male presents to your office with a seven day history of a right sided headache, located in the temporal area. He has noted that it is painful to brush his hair. He has also found that he cannot chew for long periods before he gets aching in the jaw on the same side. His shoulders and thighs ache particularly in the morning. On examination, he has tenderness of the right temporal artery, and the artery is a little tortuous, with a faint pulse. Vision appears normal. His joint examination is normal with no evidence of synovitis or effusions. Muscle strength is normal, and there is pain on range of motion of the shoulders and hips.

Pertinent laboratory results: ESR is 94 mm/hr and the CBC is normal.

10 x magnification of Temporal Artery Biopsy demonstrates a transmural inflammatory infiltrate with compression of the arterial lumen.

1. What is the most likely diagnosis?

2. What is the most appropriate treatment?

3. What is the differential diagnosis?

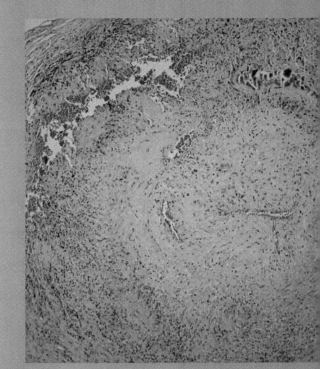

Figure 23. *Temporal Artery biopsy.*

Answers to Case 18

Diagnosis

The above patient presents with classic symptoms of Temporal Arteritis, with headaches, scalp tenderness and jaw claudication. He also has symptoms consistent with associated polymyalgia rheumatica (PMR) with morning stiffness, shoulder and thigh pain. A high ESR may be seen in both conditions.

The classic histologic picture of active temporal arteritis is a transmural inflammatory infiltrate containing multinucleated giant cells and admixed acute and chronic inflammatory cells. Associated is partial destruction of the vessel wall and disruption of the elastic lamina, fragments of which can often be identified within the giant cells.

Treatment

It is very important to start oral prednisone as soon as the diagnosis is suspected to prevent visual loss from ischemic complications. The usual dose is (1-2mg/kg) 40-60mg for temporal arteritis. If PMR is seen alone, starting dose is usually 10-20mg, and the prompt response to low dose steroids is important for the diagnosis of PMR. The dose is gradually tapered once symptoms are controlled to limit side-effects of long term therapy.

Differential Diagnosis

Tension headache is a more common cause of headache, but pain is usually felt in the frontal area, often associated with stress. Migraines may also mimic Temporal Arteritis with visual symptoms, but there is usually a previous history and the ESR is not elevated.

Case 19

This 68-year-old African-American man has had a long history of progressive pain, swelling, and instability of the knee. On exam there is marked bony enlargement, crepitus and a moderate effusion.

Knee AP and lateral radiographs are shown in Figures 24 and 25

1. **What are the findings?**

2. **What is the diagnosis?**

3. **What test might give a clue to the etiology of his problem?**

4. **What other diseases lead to this condition?**

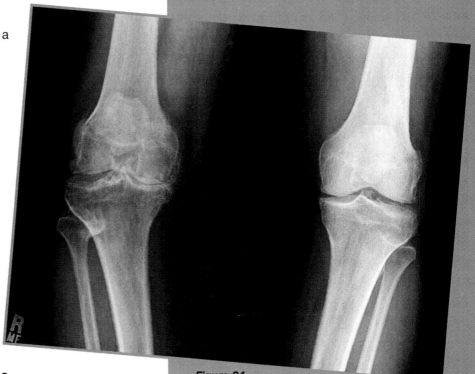

Figure 24.

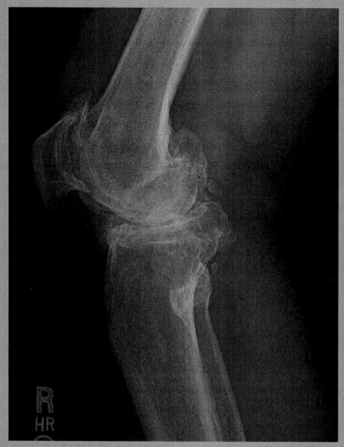

Figure 25.

Answers to Case 19

Findings

The radiographic changes show both marked eburnation of bone with exuberant spur formation as well as fragmentation of bone.

Diagnosis

This patient has Charcot (neuropathic) arthropathy due to tabes dorsalis, with impairment of pain and proprioception of the joint.

Tests suggested

This man had positive syphilis serology. Although neuropathic arthropathy is thought to be relatively painless, in late stages pain may be prominent, but somewhat less than expected with the marked clinical and radiographic findings.

In tabes dorsalis, joints typically affected include knees, hips and ankles. The radiographic findings in the knee are similar to, but much more exaggerated than the findings of degenerative joint disease. This has been referred to as "osteoarthritis with a vengeance."

Other causes of this condition

Other causes of neuropathic joint disease include diabetic neuropathic arthropathy which most commonly affects the joints of the forefoot and ankle, and syringomyelia which affects the joints of the upper extremities, especially the shoulder. Rarer causes include other neuropathies, such as leprosy, and congenital indifference to pain.

Case 20

This is a 64-year-old diabetic Caucasian male who several years ago was told he had an abnormal X-ray of his cervical spine following an auto accident. He has had slow, but progressive pain and limitation of motion of his cervical spine.

Cervical Spine and Lumbar lateral radiographs are shown in Figure 26 and 27.

1. **What are the findings and what is the diagnosis?**

2. **How can this disorder be distinguished radiographically from ankylosing spondylitis?**

3. **What extra-axial features may be seen?**

4. **What diseases are associated with this condition?**

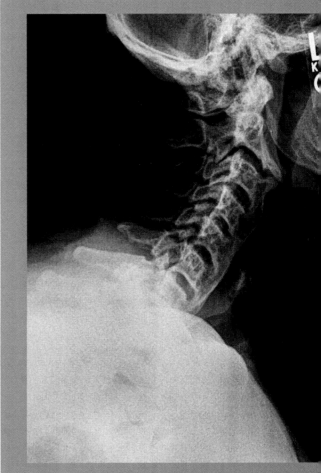

Figure 26.

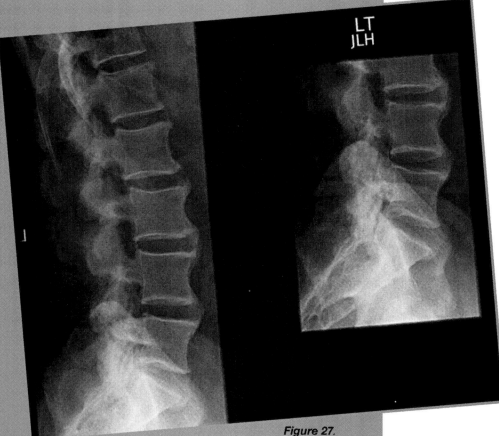

Figure 27.

Answers to Case 20

Findings and diagnosis

Cervical and lumbar X-rays reveal hyperostosis of the anterior longitudinal ligament and bony attachments of tendons and ligaments.

This patient has diffuse idiopathic skeletal hyperostosis (DISH), or Forestier's disease. The most common manifestations of DISH are in the thoracic spine, but it can also be seen in the cervical and lumbar areas. Characteristically, there is ossification of the anterior longitudinal spinal ligament. Cervical involvement may cause dysphagia.

Proposed radiographic criteria include:

1) **Presence of "flowing" calcification or ossification along the anteriolateral aspect of at least four contiguous vertebral levels**

2) **Relative preservation of the disc heights in the involved disc segments**

3) **The absence of apophyseal joint ankylosis, sacroiliac joint erosions, sclerosis, or widespread intra-articular bony ankylosis**

Distinguishing from ankylosing spondylitis

In DISH, the findings are best seen on the lateral films and the ossifications attach to the vertebral body several millimeters above and below the superior and inferior vertebral margins. In ankylosing spondylits, the syndesmophytes arise from the superior and inferior margins of the vertebral bodies and may best be seen in the frontal projections.

Extra-axial features seen in DISH

Patients often have associated osteoarthritis and enthesophytes (bony excrescences at the attachment of tendons and ligaments to bone) in peripheral joints.

Diseases are associated with DISH

DISH is associated with the metabolic syndrome and diabetes mellitus.

Case 21

This 53-year-old man presents with acute pain and decreased range of motion of the left shoulder for the past three days. He has noted the gradual onset of some pain in this shoulder for the past six months. Physical examination reveals point tenderness just below the acromial process.

1. What are the radiographic findings?

2. What is etiology of this condition?

3. What is the treatment?

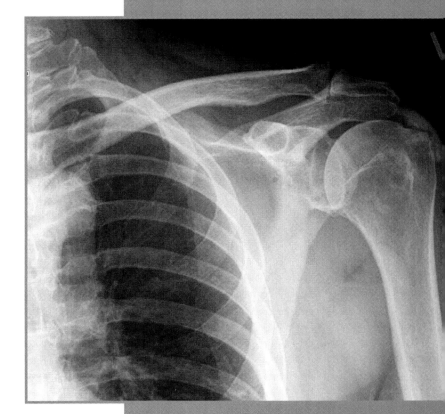

Figure 28. Left shoulder radiograph.

Answers to Case 21

Radiographic findings

The X-ray demonstrates calcific tendinitis of the shoulder. Additionally, there is a large calcium deposit in the subacromial bursa.

Etiology

The cause of calcific tendinitis, while unknown, is thought to be due to degeneration of the tendon leading to calcification through a dystrophic process. Most commonly, the supraspinatus tendon is involved and many of these patients are young and have an underlying impingement syndrome.

The classic description of this entity by Codman elaborates the natural history of this disorder. There are phases of pain, spasm and limitation of motion and eventual atrophy. The process starts as degeneration of the tendon, followed by calcification, and rupture into the subacromial bursa, which is sometimes the cause of acute symptoms (see Figure 28). Interestingly, there is poor correlation between X-ray findings and clinical symptoms. If untreated, pain and decreased range of motion may lead to adhesive capsulitis, also known as "frozen shoulder".

Treatment

Treatment of the acute phase includes non-steroidal anti-inflammatory drugs and local steroid injections. If present, an impingement should be corrected surgically.

Case 22

A 53-year-old African American man presents with a three-month history of low grade fevers, night sweats, cough, and weight loss of 30 pounds. He denies any history of foreign travel or prison time and he has no sick contacts. He is a non-smoker. He denies arthralgias, rashes, oral or genital ulcers, alopecia, numbness in the hands or feet, or muscle weakness. He admits to diffuse myalgias associated with the fevers. Two months ago, he was treated for pneumonia without improvement in his symptoms.

On exam, he is a thin male with normal lung and cardiac exam. There is no lymphadenopathy. Musculoskeletal examination reveals no synovitis. Strength and sensation are also normal.

Chest X-ray reveals enlarged hilar lymph nodes, confirmed by CT scan of the chest.

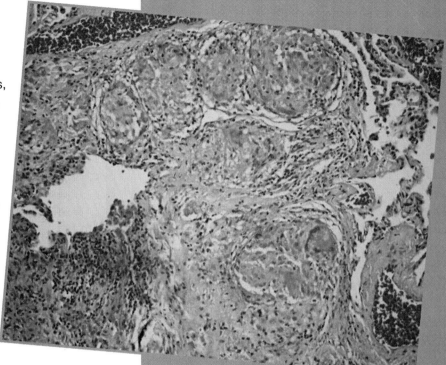

Figure 29.
Biopsy of a hilar lymph node is shown 40 xs.

1. What are the findings and what is the diagnosis?

2. What are the histopathologic features and
   differential diagnosis?

3. What are the stages of pulmonary involvement and
   how does this relate to prognosis?

4. What are treatment recommendations?

Answers to Case 22

Findings

This patient has sarcoidosis based on the pathologic findings composed of granulomas containing epithelioid histiocytes, multinucleated giant cells and a thin rim of mature lymphocytes shown in Figure 29. There is no caseation.

Histopathologic features

The characteristic histopathologic lesion of sarcoidosis is a non-necrotizing "immune" granuloma composed of epithelioid histiocytes and multinucleated histiocytes surrounded by a variable rim of lymphocytes.

The granulomas can be found in virtually every organ in the body. In the lung, they are characteristically distributed along the pleura, interlobular septae and bronchovascular bundles, corresponding to the distribution of the lymphatic channels. Although not specific for sarcoidosis, several distinctive inclusions, termed asteroid bodies and Schaumann bodies, can often be found. Non-caseating granulomas in hilar lymph nodes and peripheral lung parenchyma are the hallmark of sarcoidosis, but rarely caseating granulomas can occur with sarcoid. When caseating granulomas are present, TB and fungal infection must be ruled out.

AFB and fungal stains and cultures are typically done on each specimen and were negative in this patient. Necrotizing sarcoid granulomatosis is a rare form of sarcoid that has been documented in numerous case reports. This is distinguished from sarcoidosis by the presence of cavitation on radiographic studies and the presence of necrosis and granulomatous vasculitis on histopathology.

Stages of Pulmonary Involvement

Prognosis for sarcoidosis is based on stage:

Stage 0 - Normal chest X-ray (5% to 10%)

Stage I - Bilateral hilar and mediastinal lymphadenopathy

Stage II - Bilateral hilar or mediastinal adenopathy or both, as well as pulmonary infiltrates

Stage III - Parenchymal abnormalities; however, there is no hilar or mediastinal lymphadenopathy

Stage IV - Pulmonary fibrosis

More than 70% of patients with stage 0 to II will remit spontaneously. Less than 20% of stage III experience remission and stage IV does not remit.

Treatment

There are no randomized controlled trials involving treatment of sarcoidosis. Historically, treatment is typically given to patients with life-threatening or organ threatening disease: severe pulmonary involvement that is life threatening, hypercalcemia, cardiac involvement, neurologic or eye involvement. Treatment is typically oral steroids. If steroids are needed for a prolonged period of time, glucocorticoid sparing agents such as hydroxychloroquine, methotrexate, or azathioprine are often used.

Case 23

A 56-year-old female presents with a three year history of a symmetric polyarthritis, intermittent bouts of episcleritis, and intermittent abdominal pain. She has had only marginal improvement of her arthritis with methotrexate (up to 20 mg orally weekly) and prednisone (15 mg daily). For the past seven months she reports painful erythematous wheals, lasting 24 to 72 hours (see Figure 30).

Laboratory results

WBC, 12,000/mm³ (neutrophils, 75%, lymphocytes, 20%, mononuclear phagocytes, 5%) Hgb, 11.3 g/dL, platelets, 220,000/mm³. ESR, 75 mm/h

AST, 55, ALT, 60, alkaline phosphotase, 90, BUN, 12 Cr, 1.0. ANA speckled pattern, 1:160, Negative auto antibodies: dsDNA, SSA, SSB, Smith and, RNP antibodies, Negative RF, and CCP antibodies. Cryoglogulins were not detected. Hepatitis B surface Ag and Hepatitis C antibodies were also negative as were C- and P- ANCAs. SPEP showed a polyclonal hypergammaglobulinemia consistent with chronic inflammation.

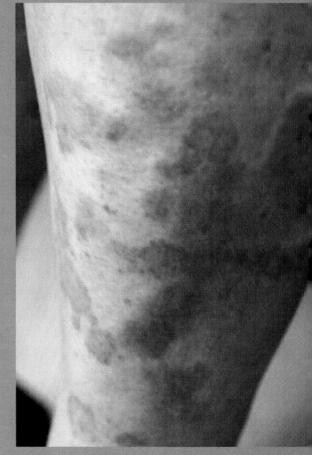

Figure 30.

Complement levels were low: C3: 20 mg/dl
(60-175 mg/dL); C4, 3 mg/dl (15-40mg/dL); CH50,
<13 (31-66 U /ml); C1Q, undetected (5-8.6 mg/dL).
Urinalysis - normal.

1. **What are the skin findings in this case?**

2. **What is the diagnosis?**

3. **How are the cutaneous lesions of this condition**
   different from common urticaria?

4. **How do we establish the diagnosis in this case?**

Answers to Case 23

Skin Findings

This patient has annular and semiannular urticarial patches. Resolution of the wheals can be associated with petechiae and purpura as in this case. Additional findings of this condition not demonstrated here include ecchymoses. Wheals can have a central dark red or brown macule, signifying underlying purpura and vasculitis. Angioedema has been reported in up to 40% of cases.

Diagnosis

This patient has hypocomplementemic urticarial vasculitis syndrome (HUVS). This is an autoimmune disorder involving six or more months of urticaria with hypocomplementemia, in the presence of various systemic findings including arthritis or arthralgias, mild glomerulonephritis, uveitis or episcleritis and recurrent abdominal pain. Criteria have been proposed for the diagnosis of HUVS. When there are insufficient systemic symptoms to meet criteria for HUVS, the diagnosis is hypocomplementemic urticarial vasculitis.

This patient has additional findings of episcleritis, inflammatory arthritis, and abdominal pain, and thus fulfills the proposed criteria for HUVS. Additional systemic manifestations of HUVS include Jaccoud's arthropathy, COPD, asthma, pleuritis, pleural effusions, nausea, vomiting and diarrhea. Ophthalmolgical disease is less commonly associated with HUV, and includes episcleritis, conjunctivitis, and uveitis.

Cardiac disease is rarely associated with HUV, and includes pericarditis, cardiac tamponade, and valvular disease. Central nervous system involvement includes aseptic meningitis, cranial nerve palsies, peripheral neuropathies, and pseudotumor cerebri.

Hypocomplementemia is a marker for systemic disease. The most commonly depressed complement components include C1Q, C3 and C4, and CH50 may be low due to activation of the classical complement pathway.

Cutaneous lesions

Urticarial vasculitic (UV) lesions can be painful and pruritic in up to one third of patients, whereas common urticarial lesions are pruritic only. The lesions of urticarial vasculitis persist for up to 72 hours, compared with common urticarial lesions that typically last from four to 36 hours.

Establishing the Diagnosis

Biopsy of the urticarial lesions is an important step in establishing the diagnosis and reveals leukocytoclastic vasculitis (LCV) of the small vessels, largely involving the postcapillary venules. Immunofluorescence reveals deposits of immunoglobulins, complement, or fibrin around blood vessels in most patients with UV.

Case 24

The patient is a 40-year-old female with progressive systemic sclerosis (PSS) who reports pain and discomfort of the buttocks for several months. She reports tenderness, nodularity and induration of the skin. The patient has a diagnosis of PSS for the past 15 years with manifestations including Raynaud's phenomenon with digital ulcerations, inflammatory myopathy, interstitial pulmonary fibrosis, and mild pulmonary hypertension.

Physical examination reveals the patient to be in moderate discomfort due to buttock pain while seated. Skin findings are remarkable for sclerodactyly, digital pitting scars, and several crusted ulcers over the distal index fingers. There is diffuse skin tightening involving trunk and extremities. The buttocks are indurated with a nodularity to the skin; there are no ulcerations.

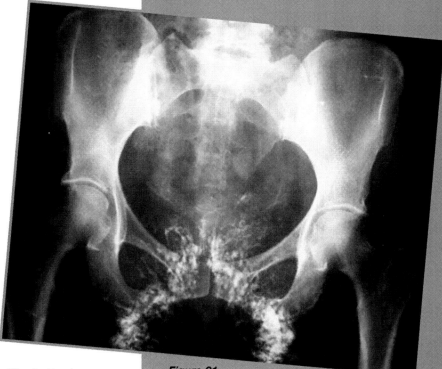

Figure 31.
Pelvic X-ray.

117

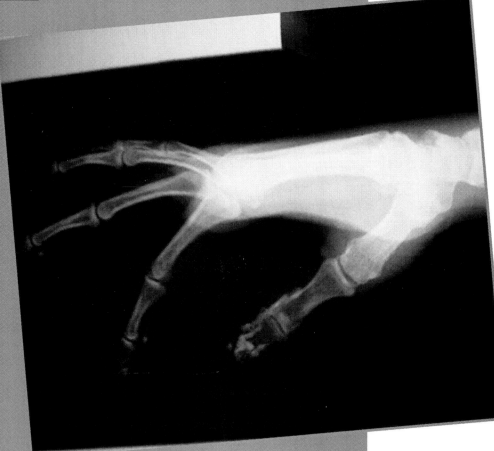

Figure 32. Lateral Hand X-ray.

1. What are the findings?
2. What is the diagnosis?
3. What is the treatment?

Answers to Case 24

Findings

There are extensive subcutaneous calcifications in the buttocks and distal first through third digits.

Diagnosis

This patient has calcinosis cutis. This complication occurs in approximately 40% of patients with long-standing PSS, and more commonly in limited disease. Calcinosis also occurs in dermatomyositis, and can often be extensive involving both the muscles and subcutaneous tissue. In PSS, the most characteristic locations of calcinosis cutis include the fingers, olecranon and prepatellar bursae, ischial tuberosities, and lateral malleoli. These deposits may become inflamed resulting in significant discomfort as in this case. They can also become superinfected, and may drain spontaneously.

Treatment

Calcinosis can be very debilitating to the patient, and unfortunately there are no proven effective therapies to dissolve or prevent progression. Various therapies have been tried and failed including warfarin, probenecid, and colchicine. Diltiazam seemed to demonstrate improvement in the calcinosis in a series of four patients. When the lesions become inflamed and painful, NSAIDS might be useful. Antibiotics, local therapy and possible removal are necessary when the lesions become superinfected.

Case 25

The patient is a 30-year-old woman with a four month history of painless lesions over the knuckles, and elbows. She additionally reports a 15 pound weight loss, decreased appetite and proximal muscle weakness. She has noticed difficulty arising from a chair and combing her hair due to the weakness.

Physical examination of the skin reveals the abnormalities as demonstrated in Figure 33.

There is erythema in a V-neck distribution on the upper back and chest.

Neurological examination is remarkable for proximal muscle strength 3/5 (active movement against gravity) lower extremities, 4/5 (active movement against gravity and some resistance) upper extremities. Distal strength and sensory examination are within normal limits.

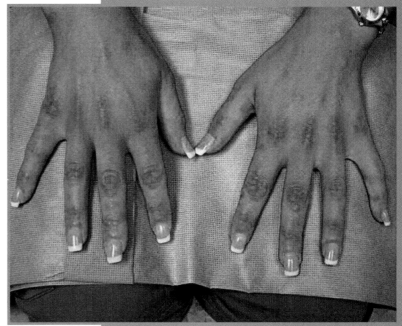

Figure 33.

Laboratory

CPK 2500 U/L(normal nl 30-70 U/L); ANA speckled
pattern 1:640; ESR 65 mm/hr.

1. What are the characteristics of the lesions
    demonstrated? What are they called? What other
    skin findings may be present in this disease?

2. What is the underlying diagnosis?

3. What further tests should be pursued?

Answers to Case 25

Characteristics of the lesions

The lesions are purplish red papules located over the PIPs, MCPs and DIPs. These are known as Gottren's papules. These lesions may also involve the skin over the olecranon process as in this case, malleoli, and patellae. They may evolve into atrophic plaques with pigmentary alterations and telangiectasias. Although not shown on figure 33, this patient also has the characteristic shawl sign rash in a "V" distribution over the upper chest and back. This rash is photosensitive, and associated with the anti-Mi2 Ab. Additional skin abnormalities, not found in our patient, include a heliotrope rash (a lilac suffusion) of the eyelids, and abnormal capillaries of the nailfold, demonstrating drop out and corkscrew dilatation of these capillaries. Up to one third of patients will have scaling and cracking of the radial aspects of the fingers termed "mechanic's hands." This finding is associated with anti-Jo-1 antibodies and interstitial lung disease, and is termed the anti-synthetase syndrome.

Underlying diagnosis

This patient has dermatomyositis as manifested by the classic shawl sign, erythematous rash in a "V" distribution over the neck and upper chest, proximal muscle weakness and elevated CPK.

Further tests

The literature supports an association of dermatomyositis and malignancy. The most associated cancers are those common to the particular patient with respect to age and sex. A thorough physical, history and laboratory examination in addition to age appropriate cancer screening is indicated in patients presenting with this diagnosis. Additional testing including CT abdomen, chest, and pelvis can be considered in patients at high risk of malignancy. For women, a transvaginal ultrasound scan and CA125 are important for earlier identification of ovarian cancer, which is overrepresented in dermatomyositis.

Case 26

A 47-year-old male with a history of bone marrow transplantation four years ago for Waldenstrom macroglobulinemia is referred to rheumatology for evaluation of skin changes. He presents with a diagnosis of scleroderma by dermatology. He denies any fevers, night sweats, weight loss, or rashes. There was no dysphagia, heartburn, dry cough, Raynaud's phenomenon, or oral ulcers. Past medical history was otherwise unremarkable.

Examination reveals thickened skin over both forearms and the posterior aspect of his neck with no involvement of the hands or other areas of his body. There was no sclerodactyly, telangiectasias, or calcinosis noted. Cardiac and pulmonary exam are normal.

1. What is the diagnosis?

2. What is the prognosis?

3. What is the best treatment?

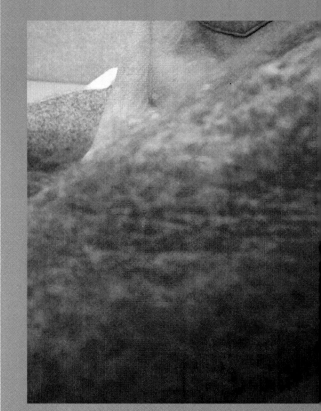

Figure 34. Skin changes of the left upper arm.

Answers to Case 26

Diagnosis

This patient has the characteristic skin findings of graft verses host disease (GVHD). GVHD occurs in 30% to 60% of patients after allogenic bone marrow transplantation. Chronic GVHD usually appears between 50 and 200 days after transplantation. While lichenoid skin changes are more common, scleroderma skin changes can occur.

Unlike primary scleroderma, GVHD associated scleroderma typically spares the fingers, is very limited, is not associated with Raynaud's phenomenon, and does not involve internal organs. Autoantibody production such as antinuclear antibodies, anti-erythrocyte antibodies, rheumatoid factor, and circulating anticoagulants are frequently seen in chronic GVHD but not in acute GVHD.

Prognosis

The prognosis of GVHD depends on its severity. Progressive onset, lichenoid skin changes, elevated serum bilirubin, persistently low platelets, or failure to respond to nine months of therapy are all poor prognostic indicators. If two or more of these are present, the mortality rate can be up to 80%. If none of these are present, the mortality rate is 30%.

Treatment

There is no specific treatment for the scleroderma-like cutaneous manifestations associated with GVHD.

Case 27

A 65-year-old man presents with progressive fatigue, fevers, weight loss of ten pounds, morning stiffness diffusely, and difficulty with daily tasks. He states that his symptoms began two to three months ago with fatigue and low-grade fevers. He cut his hair very short and has grown a beard and has difficulty combing his hair and shaving due to arm weakness. On examination, he does not have any rashes, oral ulcers, or hair loss. There is some tenderness over his proximal muscles; there is no synovitis. Cardiac exam is normal, but fine crackles are heard at bilateral lung bases. His muscle strength is 3/5 in the proximal muscles of the upper and lower extremities, and 4/5 in the distal muscles of the upper and lower extremities. The laboratory results are as below:

WBC, 9,000
Hgb 11 (normal 12-17 g/dl)
CPK 2,450 units (normal 30-170 U/L)

1. **What are the pathological findings and what is the diagnosis?**

2. **What are the clinical features?**

3. **Is further testing required?**

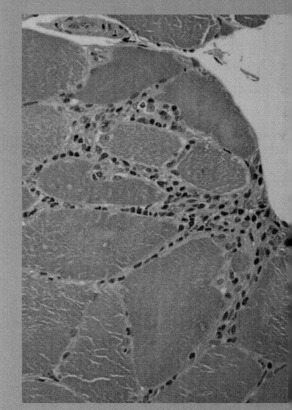

Figure 35. Muscle biopsy.

Answers to Case 27

Pathological findings

This muscle biopsy demonstrates endomysial inflammation (surrounding the myocytes) compatible with a diagnosis of polymyositis. The histologic features of polymyositis are non-specific and can mimic other inflammatory myopathies. Typically, an endomysial inflammatory infiltrate composed of mononuclear cells is identified around and within viable myocytes. Scattered foci of necrotic and regenerating myofibers may also be seen.

In contrast to the above findings seen in polymyositis the location of inflammation seen in dermatomyositis involves the muscle fascicle. T cell infiltrate is usually present in polymyositis whereas a B cell infiltrate is predominant in dermatomyositis.

Clinical features

Polymyositis (PM) is an inflammatory myopathy that causes weakness of the proximal muscles of the arms and legs as well as the neck. Additional features include fatigue, fever, and weight loss. EMG is used along with muscle biopsy for diagnosis, but 10% to 15% of patients with PM can have a normal EMG. Elevated serum skeletal muscle enzymes are characteristic of polymyositis and include CPK (creatine phosphokinase), aldolase, AST and ALT (aspartate and alanine aminotransferases) and LDH (lactate dehyrogenase). CPK is the most consistently abnormal, however, a normal CPK may be seen in 15% of patients with polymyositis and in this situation, measurement of the other enzymes may improve diagnostic accuracy.

Further Testing

Polymyositis can involve other organs such as the lungs and heart. High resolution CT scan of the chest and pulmonary function tests should be done to check for signs of interstitial lung disease (ILD). The presence of anti-Jo-1 antibodies is associated with an increased prevalence of ILD. EKG and echo should be performed to check for cardiac involvement. Any patient diagnosed with myositis should undergo age-appropriate cancer screening.

Visual Diagnosis Book Abbreviations

AS	Ankylosing Spondylitis
OA	Osteoarthritis
RV	Rheumatoid Vasculitis
RA	Rheumatoid Arthritis
PAN	Polyarteritis Nodosa
SLE	Systemic Lupus erythematosus
DIP	Distal interphalangeal (joints)
PIP	Proximal interphalangeal (joints)
MTP	Meta-tarsal phalangeal (joints)
MCP	Meta-carpel phalangeal (joints)
NSAID	Non-steroidal anti-inflammatory medications
DMARDS	Disease Modifying anti-rheumatic medications
TNF	Tumor Necrosis Factor
ESR	Erythrocyte sedimentation rate
CRP	C reactive protein
ANA	Anti-nuclear antibody

ANCA	Anti-neutrophilic cytoplasmic antibody
MRI	Magnetic Resonance Imaging
EMG/NCS	Electromyography and Nerve Conduction Study

Reference Ranges

Hematology

Complete Blood Count (CBC)

Hemoglobin	(Hb)	12-17 g/dl
Heamtocrit		36-50%
White Blood	(WBC)	4,500-11,000/mm^3
Differential		
	Basophils	0-3%
	Eosinophils	0-7%
	Lymphocytes	20-50%
	Mononuclear cells	2-12%
	Neutrophils	42-75%
Platelets		150,000-400,000
ESR		0-20 mm/h (male)
		0-30 mm/h (female)

Chemistry

Basic Metabolic Profile

BUN	10-20 mEq/L
Calcium	8-10mg/dL
Creatinine	0.6-1.4 mg/dL

Hepatic function

Albumin	3.5-5.0g/dl
Alkaline Phosphotase	39-117 U/L
ALT	8-50 U/L
AST	15-46

Other

CK	30-170 U/L
Uric acid	4.4-7.6 mg/dL

Auto-antibodies and Immunologic testing

Anti-cardiolipin (aCL) antibodies

IgG	<13 U/ml
IgM	<17 U/ml
IgA	<20 U/ml

Antinuclear Ab (ANA)	<1:40	
Anti-CCP antibodies	<14 IU/ml	
Anti-centromere antibodies	Negative	
Anti-double stranded dna antibody	<30 IU/ml	negative
	30-50 IU/ml	borderline
	>50 IU/ml	positive
Anti-Jo-1 antibodies	<1.0 EU/ml	
Anti-RNP antibodies	<20 EU/ml	
Anti-Scl-70 antibodies	<20 EU/ml	
Anti-Smith antibodies	<20 EU/ml	
Anti-SSA (Ro antibodies	<20 EU/ml	
Anti-SSB (La) antibodies	<20 EU/ml	

ANCA

c-ANCA	negative
p-ANCA	negative
Anti-PR3 antibodies	<6 EU/ml
Anti-MPO antibodies	<7 EU /ml
RF	0-39 IU/ml
C3	60-175 mg/dL
C4	15-40mg/dL

Urine

Protein	<1+
Erythrocytes	0-2 cells/hpf
Leukocytes	0-3 cells/hpf

References

CASE 1

Dougadas M, Van Der linden S, Juhlin R et al. The European Spondyloarthropathy Study Group preliminary criteria for the classification of Spondyloarthropathy. Arthritis Rheum. 1991;34:1218-1227.

Pradeep DJ, Keat A, Gaffney K. Predicting outcome in Ankylosing Spondylitis Rheumatology. 2008;47:942-945.

CASE 2

Scott DG, Bacon PA, Tribe CR. Systemic rheumatoid vasculitis: a clinical and laboratory study of 50 cases. Medicine (Baltimore). 1981; 60:288.

Voskuyl AE, Zwinderman AH, Westedt ML, et al. Factors associated with the development of vasculitis in rheumatoid arthritis: results of a case-control study. Ann Rheum Dis. 1996; 55:190.

Sayah A, English JC. Rheumatoid arthritis: A review of the cutaneous manifestations. J Am Acad Dermatol. 2005; 53:191.

Puechal X, Said G, Hilliquin P, et al. Peripheral neuropathy with necrotizing vasculitis in rheumatoid arthritis. A clinicopathologic and prognostic study of thirty-two patients. Arthritis Rheum. 1995; 38:1618.

Said G, Lacroix C. Primary and secondary vasculitic neuropathy. J Neurol 2005; 252:633.

Schmid FR, Cooper NS, Ziff M, et al. Arteritis in rheumatoid arthritis. Am J Med. 1961; 30:56.

Okhravi N, Odufuwa B, McCluskey, P Lightman, S Scleritis. Surv Ophthalmol. 2005; 50:351.

Squirrell DM, Winfield J, Amos RS. Peripheral ulcerative keratitis 'corneal melt' and rheumatoid arthritis: a case series. Rheumatology. (Oxford) 1999; 38:1245.

Sokoloff L. The heart in rheumatoid arthritis. Am Heart J 1953; 45:635.

Cruickshank B. The arteritis of rheumatoid arthritis. Ann Rheum Dis. 1954; 13:136.

van Albada-Kuipers, GA Bruijn, JA, Westedt ML, et al. Coronary arteritis complicating rheumatoid arthritis. Ann Rheum Dis. 1986; 45:963.

Johnson, RL, Smyth, CJ, Holt, GW, et al. Steroid therapy and vascular lesions in rheumatoid arthritis. Arthritis Rheum 1959; 2:224.

Harper, L, Cockwell, P, Howie, AJ, et al. Focal segmental necrotizing glomerulonephritis in rheumatoid arthritis. QJM 1997; 90:125.

Conn, DL, Schroeter, AL, McDuffie, FC. Immunopathologic study of sural nerves in rheumatoid arthritis. Arthritis Rheum 1972; 15:135.

CASE 3

Terkeltaub RA. Clinical Practice Gout N Engl J Med. 2003;349:1647-1655.

CASE 4

Lightfoot, RW, Michet, BA, Bloch, DA, et al. *The American College of Rheumatology 1990 criteria for the classification of polyarteritis nodosa.* Arthritis Rheum 1990; 33:1088.

Balow, JE. *Renal vasculitis.* Kidney Int 1985; 27:954.

Jennette, JC, Falk, RJ. *The pathology of vasculitis involving the kidney.* Am J Kidney Dis 1994; 24:130.

CASE 5

JH Klippel :Primer on the Rheumatic Diseases 13th Ed. 2008 McCarthy G, *Calcium Pyrophosohate Dihydrate, Hydroxyapatite, and Miscellaneous crystals.* Chapter 13:263-270.

CASE 6

Fabbri, P, Cardinali, C, Giomi, B, Caproni, M. *Cutaneous lupus erythematosus: Diagnosis and management.* Am J Clin Dermatol 2003; 4:449.

Tuffanelli, DL. *Bullous eruption in lupus erythematosus.* Ann Intern Med 1983; 98:261.

Cutaneous immunopathology: Recent observations. J Invest Dermatol 1975; 65:143.

Gammon, WR, Briggaman, RA. *Bullous, SLE: A phenotypically distinctive but immunologically heterogeneous bullous disorder.* J Invest Dermatol 1993; 100:28S.

Gandhi, K, Chen, M, Aasi, S, et al. *Autoantibodies to type VII collagen have heterogeneous subclass and light chain compositions and their complement-activating capacities do not correlate with the inflammatory clinical phenotype.* J Clin Immunol 2000; 20:416.

CASE 7

Barrow M.V., Holubar K.: *Multicentric reticulohistiocytosis. A review of 33 patients.* Medicine (Baltimore) 48. 287-305.1969.

Yee K.C., Bowker C.M., Tan C.Y., et al: *Cardiac and systemic complications in multicentric reticulohistiocytosis.* Clin Exp Dermatol 18. 555-558.1993.

Lambert C.M., Nuki G.: *Multicentric reticulohistiocytosis with arthritis and cardiac infiltration: regression following treatment for underlying malignancy.* Ann Rheum Dis 51. 815-817.1992.

Bauer A., Garbe C., Detmar M., et al: *[Multicentric reticulohistiocytosis and myelodysplastic syndrome].* Hautarzt 45. 91-96.1994;[in German].

Catterall M.D., White J.E.: *Multicentric reticulohistiocytosis and malignant disease.* Br J Dermatol 98. 221-224.1978.

Oliver G.F., Umbert I., Winkelmann R.K., et al: *Reticulohistiocytoma cutis – review of 15 cases and an association with systemic vasculitis in two cases.* Clin Exp Dermatol 15. 1-6.1990.

Snow J.L., Muller S.A.: *Malignancy-associated multicentric reticulohistiocytosis: a clinical,histological and immunophenotypic study.* Br J Dermatol 133. 71-76.1995.

CASE 8

Gladman DD Effectiveness of Psoriatic Arthritis Therapies. Seminars in Arthritis and Rheumatism. 2003:33:1; 29-37.

CASE 9

Saag KG et al American College of Rheumatology 2008 Recommendations for the Use of Nonbiologic and Biologic Disease-Modifying Antirheumatic Drugs in Rheumatoid Arthritis. Arthritis & Rheumatism (Arthritis Care & Research) Vol 59 No 6 June 15 2008 762-784.

CASE 10

Carter JD. Reactive Arthritis Infect Dis Clin N Am 20 (2006) 827-47.

CASE 11

Du Bois Lupus 5th Edition 1997 Cutaneous Manifestations of Lupus. Sontheimer RD, Provost T p569-623.

CASE 12

Flendrie M, Vissers WH, Creemers MC, de Jong EM, cande Kerkhof PC, van Riel PL. Dermatological conditions during TNF-alpha-blocking therapy in patients with rheumatoid arthritis: a prospective study. Arthritis Res Ther 2005;7:666-76.

Carter JD, Gerard HC, Hudson AP. Psoriasiform Lesions Induced by TNF Antagonists: A Skin-Deep Medical Conundrum. Annals of the Rheumatic Diseases 67(8):1181-3, 2008. Epub 2008 Feb 25.

Sfikakis PP, Iliopoulos A, Elezoglou A, Kittas C, Stratigos A. Psoriasis induced by anti-tumor necrosis factor therapy: a paradoxical adverse reaction. Arthritis Rheum 2005;52:2513-18.

Ritchlin C, Tausk F. A medical conundrum: onset of psoriasis in patients receiving anti-tumor necrosis factor agents. Ann Rheum Dis 2006;65:1541-4.

Camisa C. In: Handbook of Psoriasis, 2nd Edition; Blackwell Publishing, Inc 1998:21.

CASE 13

JH Klippel Primer on the Rheumatic Diseases 13th Edition. Chapter 21 Vasculitides Immune Complex Mediated Vasculitis Seo, P p.427-434.

McIlwain L, Carter JD, Bin-Sagheer S, Vasey FB, Nord J. Hypersensitivity Vasculitis with Leukocytoclastic Vasculitis Secondary to Infliximab. Journal of Clinical Gastroenterology. 36(5):411-3, 2003.

Mohan N, Edwards ET, Cupps TR et al. Leukocytoclastic vasculitis associated with Tumor necrosis factor-alpha blocking agents. J Rheum 2004;31(10)1955-8.

CASE 14

Baag AA, Vale FL, Carter JD, Rojiani AM. Granulomatosis with CNS Involvement: A Neuroimaging Clinicopathologic Correlation. Journal of Neuroimaging Epub 2008 Apr 3.

Morgan MD, Harper L, Williams J, Savage C. Anti-neutrophil cytoplasm – associated glomerulonephritis. J Am Soc Nephrol. 2006;17(5):1224–1234.

Nishino H, Rubino FA, DeRemee RA, et al. Neurological involvement in Wegener's granulomatosis: an analysis of 324 consecutive patients at the Mayo Clinic. Ann Neurol 1993;33(1):4-9.

CASE 15

Burns TM. Neurosarcoidosis. Archives of Neurology 2003:60(8)1166-8.

Bihan H, Christozova V, Dumas JL et al. Sarcoidosis: clinical, hormonal and MRI manifestations of hypothalamic pituitary disease in 9 patients and review of the literature. Medicine(Baltimore) 2007:86;259-68.

Carter JD, Bognar B, Valeriano J, Vasey FB. Refractory neurosarcoidosis: a dramatic response to Infliximab. American Journal of Medicine. 117(4):277-9, 2004.

CASE 16

Pastores GM, Meere PA. Musculoskeletal Complications associated with lysosomal storage disorders. Curr Opin Rheumatol 2005;17:70-78.

CASE 17

Edwards, CQ, Kushner, JP. Screening for hemochromatosis. N Engl J Med 1993; 328:1616.

Petrangelo A. Hereditary Hemochromatosis- A new look at an old disease. N ENgl J Med 2004;350:2383-2397.

Tavill, AS. Diagnosis and management of hemochromatosis. Hepatology 2001; 33:1321.

CASE 18

Hunder GG, Bloch DA, Michel BA et al. The American College of Rheumatology criteria for the classification of giant cell arteritis Arthritis Rheum 1990-;33:1122-1128.

CASE 19

Resnick D: Neuropathy. In Resnick D, Niwayama G (eds): Diagnosis of Bone and Joint Disorders. Philadelphia, WB Saunders, 1981, p 2430-2431.

CASE 20

Resnick D, Niwayama G: Radiographic and pathologic features of spinal involvement in diffuse idiopathic skeletal hyperostosis (DISH). Radiology 119:559, 1976.

Sarzi-Puttini P, Atzeni F. New developments in our understanding of DISH (diffuse idiopathic skeletal hyperstosis). Curr Opin Rheumatol 2004; 16.287-292.

JH Klippel. Primer on Rheumatic Diseases 13th Ed 2008 (Chapter11 Osteoarthritis Clinical Features) Paul Dieppe MD p224-228.

CASE 21

Primer on the Rheumatic Diseases 13th Ed Klippel, JH Chapter3 Musculoskeletal Signs and Symptoms Biundo JJ p68-93
Codman EA: The Shoulder: Rupture of the Supraspinatus Tendon and Other Lesions in or About the Subacromial Bursa. Boston, Thomas Todd, 1934.

CASE 22

Wasfi Y, Meehan R, Newman LS. Sarcoidosis. Chapter 105. Harris:Kelly's Textbook of Rheumatology 7th ed. Philidelphia, PA: Saunders; 2005. MD Consult Web site. Available at http://home.mdconsult.com/das/book/body/109460664-8/769078826/1257/852.html.

Wasfi Y, Newman LS. Sarcoidosis. Chapter 55. Mason: Murray & Nadel's Textbook of Respiratory Medicine, 4th ed. Philadelphia, PA: Saunders;2005. MD Consult Web site. Available at http://home.mdconsult.com/das/book/body/109460664-19/769116816/1288/499.html.

Statement on sarcoidosis: Joint statement of the American Thoracic Society (ATS), the European Respiratory Society (ERS) and the World Association of Sarcoidosis and Other Granulomatous Disorders (WASOG) adopted by the ATS Board of Directors and by the ERS Executive Committee, February 1999. Am J Respir Crit Care Med 1999; 160:736-755.

CASE 23

Kaplan, AP. Clinical practice. Chronic urticaria and angioedema. N Engl J Med 2002; 346:175.

Up to date Desk Top 16.2 Brewer Jerry D, Davis Mark DP. Urticarial Vasculitis May 2008.

McDuffie, FC, Sams, WM Jr, Maldonado, JE, et al. Hypocomplementemia with cutaneous vasculitis and arthritis. Possible immune complex s yndrome. Mayo Clin Proc 1973; 48:340.

Zeiss, CR, Burch, FX, Marder, RJ, et al. A hypocomplementemic vasculitic urticarial syndrome. Report of four new cases and definition of the disease. Am J Med 1980; 68:867.

Agnello, V, Ruddy, S, Winchester, RJ, et al. Hereditary C2 deficiency in systemic lupus erythematosus and acquired complement abnormalities in an unusual SLE-related syndrome. Birth Defects Orig Artic Ser 1975; 11:312.

Oishi, M, Takano, M, Miyachi, K, et al. A case of unusual SLE related syndrome characterized by erythema multiforme, angioneurotic edema, marked hypocomplementemia, and Clq precipitins of the low molecular weight type. Int Arch Allergy Appl Immunol 1976; 50:463.

Agnello, V, Koffler, D, Eisenberg, JW, et al. C1g precipitins in the sera of patients with systemic lupus erythematosus and other hypocomplementemic states: characterization of high and low molecular weight types. J Exp Med 1971; 134:228s.

Schwartz, HR, McDuffie, FC, Black, LF, et al. Hypocomplementemic urticarial vasculitis: association with chronic obstructive pulmonary disease. Mayo Clin Proc 1982; 57:231. Sissons, JG, Peters, DK, Williams, DG, et al. Skin lesions, angio-oedema, and hypocomplementaemia. Lancet 1974; 2:1350.

Feig, PU, Soter, NA, Yager, HM, et al. Vasculitis with urticaria, hypocomplementemia, and multiple system involvement. JAMA 1976; 236:2065.

Schultz, DR, Perez, GO, Volanakis, JE, et al. Glomerular disease in two patients with urticaria-cutaneous vasculitis and hypocomplementemia. Am J Kidney Dis 1981; 1:157.

Davis, MDP, Daoud, MS, Kirby, B, et al. Clinicopathologic correlation of hypocomplementemic and normocomplementemic urticarial vasculitis. J Am Acad Dermatol 1998; 38:899.

Mehregan, DR, Hall, MJ, Gibson, LE. Urticarial vasculitis: a histopathologic and clinical review of 72 cases. J Am Acad Dermatol 1992; 26:441.

Black, AK. Urticarial vasculitis. Clin Dermatol 1999; 17:565.

Sanchez, NP, Winkelmann, RK, Schroeter, AL, Dicken, CH. The clinical and histopathologic spectrums of urticarial vasculitis: study of forty cases. J Am Acad Dermatol 1982; 7:599.

CASE 24

Harris: Kellys Textbook of Rheumatology (Online)Chapter 79 James R Seibold.

Up to Date Online 16.2. Overview of the treatment and prognosis of scleroderma in adults.

Palmieri, GM, Sebes, JI, Aelion, JA, et al. Treatment of calcinosis with diltiazem. Arthritis Rheum 1995; 38:1646.

CASE 25

JH Klippel. Primer on the Rheumatic Diseases 13th Ed 2008 Chapter 18 Idiopathic and Inflammatory Myopathies P363-380.

Cheong WK, Hughes GR, Norris PG, Hawk JL: Cutaneous photosensitivity in dermatomyositis. Br J Dermatol 1994; 131:205.

Sigurgeirsson, B, Lindelof, B, Edhag, O, Allander, E. Risk of cancer in patients with dermatomyositis or polymyositis. N Engl J Med 1992; 326:363.

Buchbinder, R, Forbes, A, Hall, S, et al. Incidence of malignant disease in biopsy-proven inflammatory myopathy. A population-based cohort study. Ann Intern Med 2001; 134:1087.

Stockton, D, Doherty, VR, Brewster, DH. Risk of cancer in patients with dermatomyositis or polymyositis, and follow-up implications: a Scottish population-based cohort study. Br J Cancer 2001; 85:41.

CASE 26

Reddy P, Ferrara JLM. Graft-Verses-Host Disease and Graft-Verses-Leukemia Responses. Chapter 107. Hoffman: Hematology: Basic Principles and Practice, 5th ed. Philadelphia, PA: Churchill Livingstone;2008. MD Consult Web site. Available at http://home.mdconsult.com/das/book/body/122319398-4/808790226/1854/1246.html.

CASE 27

Wortmann R. Inflammatory Diseases of Muscle and Other Myopathies. Chapter 80. Harris:Kelly's Textbook of Rheumatology 7th ed. Philidelphia, PA: Saunders; 2005. MD Consult Web site. Available at http://home.mdconsult.com/das/book/body/ 109460664-9/769080859/1257/635. html.